CALIFORNIA MUSHROOMS

CALIFORNIA MUSHROOMS

A Field Guide to the Boletes

HARRY D. THIERS
San Francisco State University

HAFNER PRESS
A Division of Macmillan Publishing Co., Inc.
New York
Collier Macmillan Publishers
London

Hafner Press
A Division of Macmillan Publishing Co., Inc.
866 Third Avenue, New York, N.Y. 10022

Collier Macmillan Canada Ltd.

Library Of Congress Cataloging in Publication Data

Thiers, Harry Delbert, 1919–
California Mushrooms.

Bibliography: p.
1. Mushrooms--California--Identification.
2. Boletaceae. I. Title.
QK617.T45 589'.222'09794 74-11002
ISBN 0-02-853410-7

Printed in the United States of America

1 2 3 4 5 6 7 8 9 10–79 78 77 76 75

LIST OF MICROFICHE ILLUSTRATIONS

PREFACE

California offers a rich field for the mushroom hunter. During the rainy season, the woods abound in a wide variety of fungi, often beautiful in form and color. Some are abundant, others rare and elusive. More and more people, especially Californians, are discovering the fascination of mushrooms. Some are intrigued by their pharmaceutical properties, others are enchanted by their culinary characteristics. Yet among the numerous studies on California plants, the fleshy fungi (mushrooms in particular) have been neglected. In fact, a large number of the eighty bolete species found in California cannot be identified by any single reference now available other than this volume.

This taxonomic guide to the Boletaceae of California is an attempt to fill the gap. It is designed for both the amateur and the professional mycologist—indeed, for anyone interested in mushrooms. The species descriptions are detailed enough for the expert, yet understandable to the casual observer. General comments on habitat, and documentation of the areas where the specimens were collected, will enable the reader to make exciting discoveries of his own. Explanations of technical terms, field keys, descriptions, and illustrations should help even the beginner to identify boletes in the field without use of a microscope. Since most boletes are edible (and many are considered delicious), comments have been included on the edibility of various species.

Mycophagists (mushroom eaters) should find this guide useful in filling kitchen baskets*; nature-study teachers should find it useful in making more accurate identifications in the field or in helping their students to identify specimens; and it should enrich the experience of hikers, weekend naturalists, and visitors to our state and national parks.

A unique feature of this book is the microfiche of fifty-four full-color photographs of some of the most important species. This film card, contained in a sleeve on the inside front cover, can be used in a number of ways: in the field, collectors can simply hold it up to the sky or use a hand lens to compare specimens with the photographs. Or the microfiche can be used on a microfiche reader in any library. On the viewer the photographs are enlarged many times life size which makes more detailed examination possible. This couldn't be done if the color illustrations were in the book itself.

*The mushroom eater must be cautioned: a mushroom cannot be identified *positively* by just comparing the specimen to a color photograph, no matter how good that photograph may be.

Why a microfiche? Partly economics. But more important, film dyes give greater color fidelity than the best printers' ink. With a microfiche the reader is given more illustrations with greater fidelity at a lower cost than had they been printed in the conventional way.

Finally, this book does not pretend to record all the boletes in the State of California. The species included are based on the author's herbarium, which was begun in 1959 and now numbers over 4,000 specimens. Collecting was done in the coastal forests from Crescent City to Santa Barbara, the Cascade Range, the Klamath Mountains, and most of the Sierra Nevada; and some specimens are gathered in Riverside and San Diego counties. The eighty species described represent a substantial increase in the number of boletes reported from California prior to 1959. In addition, five extralimital species are described, on the strong possibility that they will soon be found—you are welcome to try your hand!

It is a pleasure to express my gratitude to the following: the numerous individuals who contributed collections, including my students, who, in addition, provided much of the motivation for this study; Dr. Alexander H. Smith, former Director of the University of Michigan Herbarium, who willingly shared his extensive knowledge of the boletes; my wife, who offered many helpful suggestions, prepared the Latin diagnoses, and proofread the manuscript. Thanks are also extended to the officials of the California State Parks System, the San Francisco Water Department, and Lassen and Yosemite National Parks for collecting permits in areas under their supervision. The financial support of the National Science Foundation (grants GB2760 and GB14937) and the Frederic Burk Foundation of San Francisco State University, San Francisco (grants 2053, 222, and 132), have helped to make this study possible.

CONTENTS

ABOUT BOLETES

The boletes are members of the large group of fungi characterized primarily by the formation of spores, or reproductive bodies, on a highly specialized, microscopic structure known as a basidium (plural, basidia). The presence of these basidia separates this class of fungi, the Basidiomycetes, from all others. There are two sublcasses or groups of Basidiomycetes. One of these, the Heterobasidiomycetes, is characterized by a basidium composed of two or more cells, often four, and is represented by such common fungi as the rusts, smuts, and jelly fungi. In the second subclass, the Homobasidiomycetes, the basidium is single-celled. Since the basidium of the boletes is unicellular, they belong to the Homobasidiomycetes along with the mushrooms or toadstools, polypores, and puffballs that are so commonly observed in California woodlands. The fruit body of the boletes is similar in appearance to the typical mushroom except that, in the boletes, tubes have replaced the gills, or lamellae, on the under surface of the cap. Because of the presence of these tubes and their pores, or openings, boletes are often called fleshy pore fungi.

Since the fruit body of mushrooms and boletes is a specialized structure whose major function is the production of basidia, it will hereafter be called the basidiocarp (basidium bearing body). Even the most cursory examination of the bolete basidiocarp reveals its similarity in general appearance and structure to the typical mushroom, or agaric, basidiocarp. The cap or pileus of the two is similar, and so is the stem or stipe. On some boletes, as in some gill fungi, there is a layer of tissue extending from the cap margin to the stem, thereby enclosing the tube cavity. This layer is the partial veil which, when the cap expands, is torn free from the margin of the cap, leaving a ring or collar of tissue attached to the stipe. This ring of tissue constitutes the annulus. The internal composition of the two types of basidiocarps is likewise similar. Each is composed of minute filaments woven together in such a way as to form a pseudotissue. In contrast to the dry, woody polypores, these basidiocarps characteristically have a high content of moisture and are putrescent; that is, they decompose readily after reaching maturity.

Like mushrooms, the bolete basidiocarp represents only one stage in the life cycle of the fungus. This large, fleshy structure is so intimately involved in the reproduction of these organisms that it is often considered to be the major reproductive structure of the fungus. It must be remembered that this role, however, properly belongs to the spore, more precisely called the basidiospore. The vegetative phase of most boletes and of many mushrooms exists in close association with the roots of many common forest trees, such as Douglas fir, oaks, and pines, as well as in the adjacent soil and humus. The vegetative unit is a microscopic, thread-like filament known as a hypha (plural, hyphae); collectively, the hyphae constitute the mycelium. The mycelium is often white or brightly colored and can frequently be seen adhering to the base of the stipe of the basidiocarp. The individual filaments are septate; that is, they have cross walls which divide the filament into a series of cells arranged end to end. The walls of the hyphal cells are most frequently thin; however, thick-walled hyphae are sometimes developed. The contents of the cells of the hypha, like most other plant cells, characteristically include the nucleus and cytoplasm. However, as in most fungi these structures are so small that they are extremely difficult to observe. The hyphae and mycelium of the boletes are similar in appearance and structure to those of other Basidiomycetes.

The major difference between the boletes and gill fungi, as suggested above, is that in the boletes the basidia are located on the inner surface of numerous tubes, which are typically vertically arranged on the lower surface of the pileus (except in *Gastroboletus*). These tubes, or gills in the case of mushrooms, are commonly designated as the hymenophore, or the part of the basidiocarp bearing the hymenium. The hymenium, in turn, is a layer of rather closely packed basidia plus distinctive sterile cells called cystidia. Another difference noted in the field is that, although some mushrooms grow on logs or other woody substrates, only a few boletes are found consistently on such substrates, and most occur in the soil or humus in the vicinity of woody plants.

THE BASIDIOCARP

The Pileus

Bolete pilei are typically large, often reaching 15 centimeters (cm) or more in diameter, and are rarely as small as 2–4 cm in diameter. Characteristically, they are more or less convex or bulbous in outline when young, becoming plane or planoconvex when mature. Colors range from almost black to many different shades of brown, pink, or bright red. The colors, with a few conspicuous exceptions, are relatively constant for a given species and change only slightly as the basidiocarp matures. Although there are no known white species in California, there are a few in which the color is constantly pallid. When the surface of the pileus is bruised, many boletes show some type of color change.

The nature of the surface of the pileus is of considerable taxonomic significance in the boletes. In most species of *Suillus* and in a few belonging to other genera it is viscid, or "sticky" or "slimy," to the touch. This condition results most commonly from the gelatinization of the walls of the hyphae of the cuticular or surface layer and is particularly noticeable in wet weather. In the nonviscid species the surface may be either moist or dry to the touch. The surface may vary from glabrous, or bare, to fibrillose, or covered with a layer of fibrils that may be closely appressed or loosely and irregularly arranged. In some species the arrangement of the fibrils may give a tomentose, or velvety, appearance. There are several boletes in which the fibrils become agglutinated into scales. Most commonly these scales remain closely attached to the surface, but in some boletes the tips may break free. In a few they become quite large and give

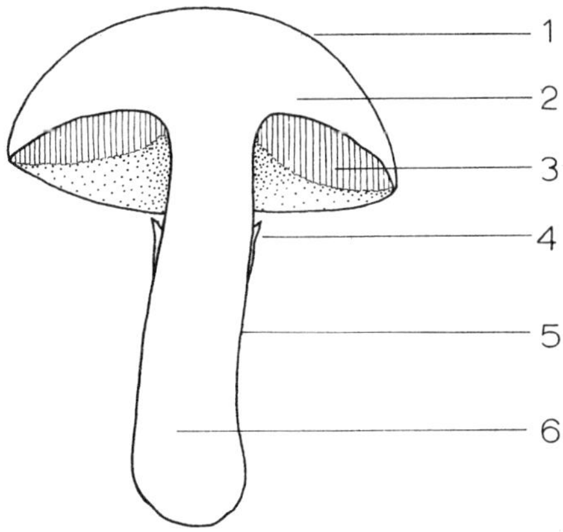

Figure I. Longitudinal section through bolete basidiocarp. 1. Pileus. 2. Pileus trama. 3. Tubes (Hymenophore). 4. Annulus. 5. Stipe cuticle. 6. Stipe trama.

the pileus a squarrose or squamulose appearance. Boletes, as they grow older, often show a tendency to become split, or rimose, on the margin and to become checked, or areolate, elsewhere on the surface. The areolations sometimes become deep and strongly pronounced in prolonged periods of dry weather, which results in a frustose condition of the pileus.

The margin or edge of the young pileus in most boletes is entire, that is, more or less smooth and even and with no ornamentations. In the genus *Suillus,* however, there is frequently a noticeable cottony roll of veil tissue closely attached to the margin of the young pilei. This roll, commonly referred to as a false veil, is typically white or whitish and usually disappears as the pileus matures. Also in the genus *Suillus* and in some other genera, a partial veil is sometimes present which, when breaking free from the pileus, often leaves fragments, or appendiculations, hanging from the margin. Typical of the members of the section *Leccinum* of the genus *Leccinum* is the continued growth of the outer layer or cuticle beyond the edge of the pileus, resulting in the formation of a band of tissue around the margin. This band usually breaks into fragments, or "flaps," as the pileus expands. These fragments seem to serve no function and eventually become either inconspicuous or completely disappear.

The flesh or context of the pileus is usually relatively soft and putrescent, and with a high water content. It varies in thickness from 0.5–2 cm, but in exceptional cases may be as thick as 4–5 cm or as thin as 2–4 millimeters (mm). Most commonly, the context is some shade of yellow or white, with other colors such as pink or tan rarely evident. The context of many boletes changes color when exposed to the air or damaged. The characteristic blue discoloration, or bluing, seen in a number of species may result from the oxidation by an enzyme of a compound known as boletol. Other color changes, such as the reddening, browning or blackening seen in many species, are probably similar oxidative reactions, but the compounds and chemical pathways are largely unknown.

The taste and odor of the context of most boletes are mild and inoffensive. *Boletus piperatus* is the only species known from California that has the peppery or acrid taste so frequently encountered in species of *Russula* and *Lactarius* of the gilled fungi. *Boletus calopus, B. rubripes,* and possibly a few others have a noticeably bitter taste, and some, such as *Suillus pungens* and *S. acerbus,* have somewhat harsh and unpleasant tastes. No boletes yet known from California have a highly distinctive odor.

The Hymenophore

The hymenophore, which consists of soft, moist, putrescent tubes, is the most distinctive feature of the boletes. The hymenium forms the inner lining of these tubes. There is a superficial resemblance between the hymenophore of polypores and boletes. The basidiocarps of polypores, however, are tough, dry, and woody.

In most boletes the hymenophore separates readily and cleanly from the pileus, and in many the tubes are easily separated from each other; however, as has been pointed out by Snell, in some cases, best exemplified by the section *Subtomentosi* of the genus *Boletus* and by the genus *Suillus,* the tubes appear to develop in a different manner and can be separated only by tearing the walls apart. Except for *Gastroboletus,* the tubes are more or less vertically arranged in an orderly fashion, and a spore print is readily obtainable. Most often the tubes are either shallowly or deeply depressed at the stipe, but in some boletes, especially species of *Suillus,* they extend down the surface in a decurrent fashion.

The pores of the tubes are worthy of special mention. Most are angular, varying from almost square to somewhat rectangular, and range in size from 0.5–2 mm in diameter. In some boletes the arrangement of the pores is radial, or boletinoid, and they radiate from the stipe much like the spokes of a wheel. Large pores compounded with internal partitions are seen in several species of *Suillus.* In a few boletes the pores are so large and elongated that they resemble lamellae and are described as lamellate. In *Boletus edulis* and related species the pores are "stuffed" when young; that is, they are filled by intrusive hyphae which with maturity of the basidiocarp collapse and disappear. The pores are usually concolorous with the tubes; however, in some species they are pink or red in contrast to the yellow color of the body of the tube or, less commonly, they may be a contrasting shade of brown or black. The pores ordinarily show the same color changes as the remainder of the hymenophore.

The Stipe

The boletes are stipitate and no sessile species are known. The stipe is central. An eccentric attachment is found rarely, except in *Gastroboletus,* and species with truly lateral stipes are still unknown. The overall shape of the stipe is sometimes equal or more or less the same size from apex to base. More commonly, it is clavate with a gradual enlargement toward the base. In some boletes, best exemplified by *Boletus satanas,* there is a conspicuous and abruptly bulbous base, which is sometimes as much as 5 cm in diameter. There are others in which the stipe is ventricose, or larger in the midportion than at the apex or base. Generally, the range in length is from 4–5 cm to a maximum of 14–15 cm, although *Suillus brevipes,* for example, may have a stipe as short as 1 cm and that of *Boletus mirabilis* may sometimes reach 20 cm. The diameter of the apex of the stipe, on an average, varies from 0.5–2 cm, but the more massive species, such as *Boletus regius* and *B. edulis,* may exceed 3–4 cm, whereas *B. piperatus, Gyroporus castaneus* and some *Suillus* species may be less than 0.5 cm in diameter.

The color of the bolete stipe ranges from white to yellow to pink to darker colors, such as red, brown, or almost black. Frequently the background color is overlain by some other pigment, resulting in a blending of colors. As in the pileus, some colors show a tendency to fade or, conversely, become more intense with age. Often the fading or darkening is uneven, which results in the formation of bands or blotches of color. Ornamentations on the surface of the stipe are often a contrasting color, and usually darken with age. Usually, the same series of color changes occurring in the pileus upon bruising or exposure takes place in the stipe, except that the changes may be more intense.

The surface of the stipe in a major percentage of boletes is either dry, moist, or, less commonly, viscid. Typically, it is glabrous, tomentose, fibrillose, or fibrillose-scaly. Sometimes it is reticulate, perhaps best demonstrated by *Boletus edulis* or *B. eastwoodiae,* in which there is a noticeable network of raised lines on the surface. A reticulum at the apical portion may be formed by the extension of the tubes down the stipe. In such cases the ridges forming the reticulum are composed of basidia and cystidia. In *Leccinum* the surface is furfuraceous, that is, covered with squamules, or scales. These are usually white or pallid when young, but change to dark brown or black with age, and are composed largely of caulocystidia, or large sterile cells. In *Suillus* the surface often has glandulae, or small raised dots, that vary considerably in size, are typically colored some shade of brown, and may stain the fingers when handled. These glandulae are composed of clusters, of fascicles, of caulocystidia. Frequently, the basal portion of the bolete stipe is clothed with coarse hairs, which may be distinctively colored. Another important taxonomic feature associated with the surface of the stipe is the annulus, or remnants of the partial veil. It is sometimes evanescent; that is, it disappears as the basidiocarp matures or is represented merely by a fibrillose zone. On the other hand, it may be massive and constitute a conspicuous part of the mature basidiocarp. In some boletes, particularly in the genus *Suillus,* the stipe may be peronate, a condition in which the lower part of the stipe is completely surrounded by velar tissue. In *Pulveroboletus ravenelii* the veil and annulus are dry, copious, brilliant yellow, and floccose in texture. In some species of *Suillus,* the annulus is white or pallid, frequently inconspicuous, and may be noticeably viscid.

The context of the stipe is typically similar in structure and texture to that of the pileus. The sequence of color changes in the stipe apex when exposed is of considerable taxonomic value in the genus *Leccinum.* The stipe is typically solid, but in a few species, such as *Gyroporus castaneus,* it is hollow, at least in the basal portion.

THE GASTROCARP

The gastroboletes have been classified in the past as puffballs rather than as boletes because they do not give a spore print. However, they are now included with the boletes because of the morphological similarity of the basidiocarps.

Because of the similarity of the gastroboletes to puffballs, different terms are used to describe the macroscopic features of the basidiocarp in the genus *Gastroboletus.* The basidiocarp is called a gastrocarp, the hymenophore is the gleba, and the stipe is the stipe-columella. The surface layer of the cap that in some species encloses the gleba is called the peridium or peridial membrane. The arrangement of the tubes in the gastrocarp is irregular; they are not vertically oriented and often radiate at various angles under the cap. The stipe-columella is often poorly developed, frequently eccentric, and often surrounded by the gleba.

THE BASIDIOSPORE

The basidiospores are of considerable diagnostic value in the taxonomy and systematics of the boletes and must be checked before positive determinations can be made. At the generic level the color of the spores in mass is of significance, and spore prints are often necessary. The spore print varies in color from a shade of brown in such genera as *Boletus, Leccinum,* and *Suillus* to yellow in *Gyroporus* and flesh or dark pink in *Tylopilus.* The individual spore ranges from hyaline or colorless to brown. The shape of the spores is, in a broad sense, more or less similar for all California species. Generally, in face view they appear elongate and cylindric to fusoid (spindle-shaped) or ellipsoid in outline. When seen in profile, however, the two sides are often unequal, with one showing a slight bulge, a condition described as ventricose. The bulge is located near the apiculate end, that is, the end attached to the basidium. Occasionally, oddly shaped or pleomorphic spores are produced, often noted when the basidium is not four-spored. Since the germination of bolete spores is exceedingly difficult, it is not known whether these atypical spores are viable or not. In *Boletus truncatus* and sometimes in other species the apical end is often abruptly terminated and appears as if it had been cut off or truncated. A germ pore or thin spot in the wall is sometimes apparent in this region.

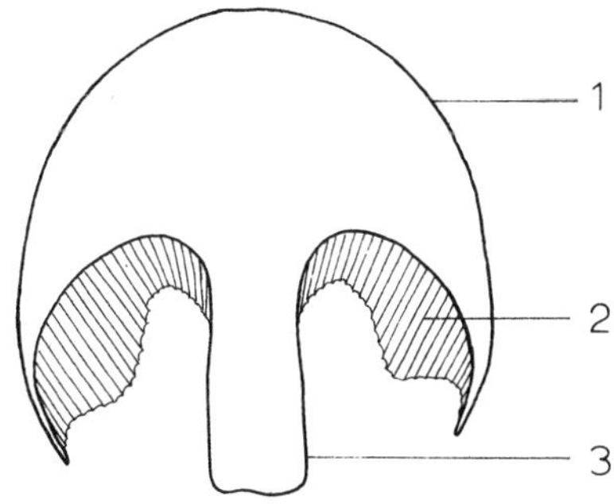

Figure II. Longitudinal section through gastrocarp. 1. Peridium. 2. Gleba. 3. Stipe-columella.

The spore length averages between 9–15 micrometers (μm) and the width from 4–5 μm. Smaller spores are characteristic of *Boletus orovillus* and several species of *Suillus.* On the other hand, spores are as long as 17–20 μm in species of *Leccinum* and in some boletes, such as *Boletus mirabilis.* "Giant" spores are sometimes seen in mounts from many different species and are presumably produced by one- or two-spored basidia. The spore wall is generally thin, averaging about 1 μm, but some spores, such as in *Boletus mirabilis,* possess thickened walls. In all known California species the outer wall appears smooth with the light microscope except that the walls of spores of *Boletus zelleri* and *B. mirabilis* have sometimes been reported as very weakly punctate or roughened.

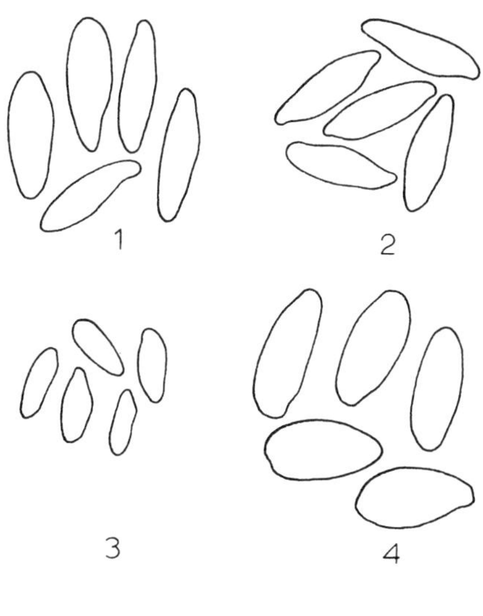

Figure III. Bolete basidiospores. 1. Boletus. 2. Leccinum. 3. Suillus. 4. Tylopilus.

The bolete spores show little or no change in color when mounted in water, dilute solutions of potassium hydroxide, or similar basic compounds. When mounted in Melzer's reagent, however, they may either remain unchanged or one of several color changes may occur. Most commonly, the brown color of the walls is merely intensified or becomes a deep rusty brown. In some the spores give a dextrinoid reaction in which they become bright rust red or tawny. The chemical reaction responsible for this color change is not fully understood, especially since rarely do all the spores show the same reaction. Recently, several species have been found in which the spore wall gives a typical blue-black, or amyloid, reaction when mounted in Melzer's reagent. Spores of *Boletus amyloideus* and *Tylopilus amylosporus* show this reaction. Like the dextrinoid reaction, the chemistry of the amyloid reaction is poorly understood, and all spores from a given species usually will not react positively.

THE HYMENIUM

The hymenium is the palisade or layer of basidia and associated cells that forms the inner lining of the tubes and, in some instances, may extend down the stipe if the surface is reticulate. The most important cells in this layer are the basidia, which develop as terminal cells of the hyphae that form the tissue of the tube. Nuclear fusion (karyogamy), reduction division (meiosis), and subsequent spore formation occur in the basidia. The basidia are rather large, club-shaped cells ranging from 20–30 μm in length and from 7–12 μm in width. At the apex four short, somewhat hooked branches develop. These are sterigmata, and eventually each will bear a single basidiospore.

Basidioles or brachybasidioles are also commonly seen in the hymenium. These may be undeveloped basidia or basidioid cells that never produce basidiospores. They appear to have the same origin as the fertile basidia, but their true function is not understood. Perhaps they serve as lateral support for the spore-bearing basidia.

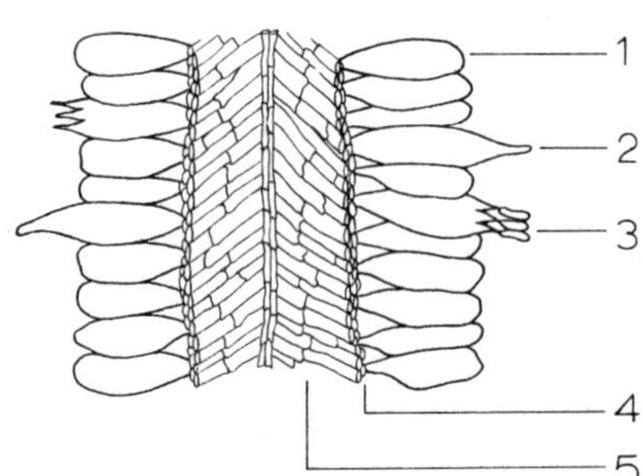

Figure IV. Longitudinal section through the bolete hymenophore. 1. Basidium. 2. Hymenial cystidium. 3. Basidium with sterigmata and basidiospores. 4. Subhymenium. 5. Bilateral trama showing central strand and divergent hyphae.

A third type of cell commonly found in the hymenium of most boletes is the cystidium. These cells are sterile and, like the basidia, arise as differentiated hyphal tips. They are often highly distinctive in size and shape, and frequently extend well beyond the basidial layer into the tube cavity. Most often they are clavate to fusoid or ventricose in shape with an elongated to obtuse or mucronate apex. Cystidia reach 50–75 μm in length and 10–15 μm in diameter. These cells are usually thin-walled, but may become noticeably thickened or incrusted with amorphous materials. When mounted in water, the cystidia are typically hyaline or only weakly pigmented; however, in potassium hydroxide or in Melzer's reagent there is often a marked color change to reddish brown or black. Most species of *Suillus* in California have at least some of the cystidia grouped together in large clusters or fascicles. These clustered cystidia when mounted in potassium hydroxide become very dark brown to black or purplish, at least in the basal portion. In the gilled fungi the cystidia occurring on the gill edge, that is, the cheilocystidia, are often quite distinct in shape and size from those on the face of the gill, the pleurocystidia. However, in most boletes there is little or no distinction between those on the pores and those on the tube lining. Therefore, they will be referred to as hymenial cystidia to distinguish them from the cystidia that sometimes occur on the stipe or pileus cuticle. The role of the cystidia is uncertain, but they may act as organs of excretion, serve as air traps, or perhaps aid in the evaporation of moisture from the hymenial surface.

THE TRAMA

The body of the pileus, hymenophore, and stipe is composed of filamentous, thin-walled hyphae that are undifferentiated, except for the surface, or cuticular, hyphae. These internal hyphae are referred to collectively as the trama. In the

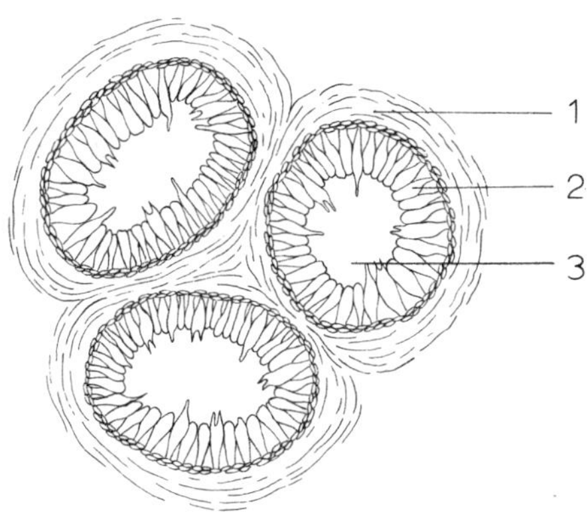

Figure. V. Crossection of bolete hymenophore. 1. Tube (hymenophoral) trama. 2. Hymenium. 3. Pore.

tubes the hyphae are characteristically arranged in a compact central strand with noticeable filaments diverging from it, except in those few boletes in which the hyphae are arranged parallel to one another. This divergent, or bilateral, trama is often apparent only in young basidiocarps and at maturity may appear tangled or interwoven. The tube trama is generally hyaline when mounted in potassium hydroxide, but may change to some shade of brown, especially in the central strand. The tramal hyphae are often relatively large and range from 4–6 μm in width. In several species the walls of the hyphae seem to gelatinize or dissolve when mounted in water or weak base solution (KOH). Occasionally, thick-walled laticiferous hyphae are irregularly distributed in the trama. They may be secretory hyphae similar to those in *Lactarius*, but no latex-bearing species of boletes are known from California except for those which secrete droplets of resinous material on the pores when young.

The pileus trama is generally similar in all bolete species and of little taxonomic or phylogenetic value. Its typical appearance is one of tangled hyphae, which are more or less loosely interwoven and usually homogeneous. Laticiferous hyphae are sometimes irregularly interspersed, but are not seen as frequently as in the tube trama. The stipe trama in all boletes is composed of masses of more or less parallel hyphae that may be loosely or tightly packed, except in *Gyroporus* where the hyphae are transversely arranged.

THE CUTICLE

One of the most important and useful anatomical features from a taxonomic point of view in the boletes is the microscopic structure of the cuticular or external layer of the pileus, and to a lesser extent of the stipe. The cuticle is usually easily distinguished from the pileus trama. In some California boletes a second layer under the outer layer, called the subcutis or hypodermium, is present; it is characterized by rather tightly interwoven hyphae and often a distinctive staining reaction in KOH and Melzer's reagent.

The simplest and perhaps most primitive type of cuticle is the cutis in which there is a layer of appressed, filamentous hyphae over the surface of the pileus. The cells of these hyphae are more or less similar in shape and size and seldom break free from one another. Such a cuticle is only rarely seen in the boletes.

More characteristic of the boletes is the trichodermial type of cuticle, which is composed of interwoven hyphae or hyphal tips that may be arranged in an upright palisade or turf. However, there is considerable variation in the trichodermial cuticle, and noticeable changes may occur as the basidiocarp matures. If at all possible, young specimens should be checked, for the hyphae of a tangled or interwoven trichodermium may collapse with age in some boletes and give the appearance of a cutis. In others the hyphal tips may remain more or less erect and highly differentiated, forming a hymeniform cuticle. If the hyphal

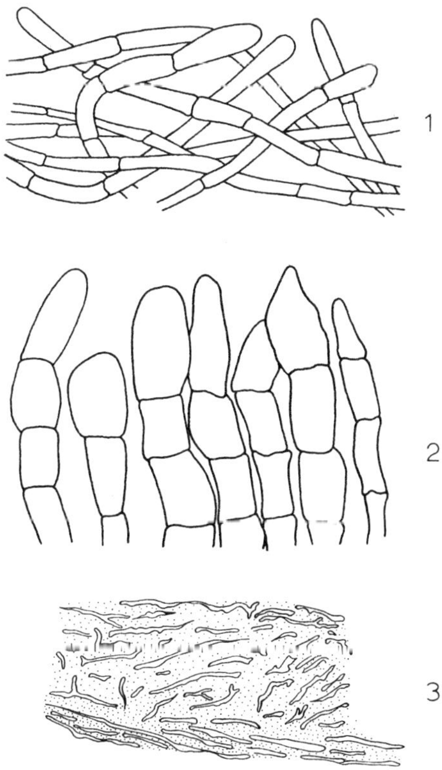

Figure VI. Types of bolete cuticle. 1. Trichodermium composed of interwoven hyphae and hyphal tips. 2. Trichodermium composed of erect hyphal tips. 3. Ixotrichodermium in which the walls of the hyphae are gelatinizing.

tips become clustered or aggregated, the surface of the cap appears tomentose or scaly. Another variation in the trichodermium occurs in many species of *Suillus* and occasionally in other genera: the walls of the hyphae gelatinize or partially dissolve; the resulting slimy or viscid layer is referred to as a pellicle or an ixotrichodermium.

In many species of *Leccinum* the individual cells of the cuticular hyphae are quite large, loosely held together, and may disarticulate or separate from one another. The walls of these cells are sometimes faintly roughened or asperulate, but this feature is usually visible only at rather high magnifications. The pigments in the cells of the cuticular hyphae of some species of *Leccinum* become aggregated into noticeable globules when mounted in Melzer's reagent. The chemistry of this reaction is not understood and neither is its significance in establishing species relationships within the genus; however, since the reaction is constant for a given taxon, it is used as a taxonomic character. In *Boletus chrysenteron* and *B. zelleri* the external surface of the cuticular hyphae become noticeably incrusted. These incrustations may be irregularly dispersed over the surface or arranged in the form of spirals. Often they become rust brown when mounted in potassium hydroxide or Melzer's reagent. Pileocystidia or sterile, highly differentiated cells somewhat similar to the cystidia in the hymenium are very rarely present in the cuticle of the boletes.

The cuticle of the stipe shows less variation than the pileus cuticle. Most commonly, the surface is differentiated as a cutis, but trichodermial or ixotrichodermial types also occur. In *Suillus* the surface of the stipe of many species is noticeably dotted with differently colored, often irregularly shaped, resinous globules. Because of their appearance, these globules are designated as glandulae; however, structurally they are composed of masses of caulocystidia similar to the cystidia in the hymenium and give similar color changes when mounted in potassium hydroxide.

Clamp connections are short, inconspicuous branches located at the cross walls or septa of the hyphae and form a by-pass around the septations. They are present in some of the boletes and are most readily found in the cuticular hyphae, at the base of the basidium or, less frequently, in the basal tomentum of the stipe. The significance of clamp connections is not fully understood since they are often not present at every septation. Their significance has been further confused by the recent observation that the vegetative hyphae of at least some species of *Suillus* when grown in pure culture develop clamps, but in the basidiocarps of the same species all hyphae are devoid of such structures. In the California flora, clamp connections have been found consistently only in *Gyroporus castaneus.*

MYCORRHIZAL ASSOCIATIONS

In California the boletes are exclusively forest-inhabiting fungi and none are known that fruit only in fields or open meadows. The restriction to the forests is due to the necessity of the formation of mycorrhizal associations in order to survive. This type of association is an intimate relationship between the vegetative mycelium of the fungus and the young roots of the associated tree. The mycelium forms a rather dense layer or mantle on the external surface of the root. This relationship seems to be mutually beneficial. The mushroom profits by obtaining nutrients and perhaps water from the host tree. The benefits afforded to the tree by the fungus are not so readily apparent, but indications are that the fungus accumulates considerable quantities of certain minerals from the soil that are subsequently available to the tree and increase the absorptive surface of the roots. Obviously, if such associations are essential for the survival or development of the tree, these fungi assume considerable significance in reforestation practices.

Since the fungus hyphae are microscopic except in mass, the difficulty of tracing the connection between the basidiocarp and the roots of the tree is

Some California Boletes and Their Probable Mycorrhizal Host

Species	Mycorrhizal Host
Boletus abieticola	Firs
Boletus aereus	Oaks
Boletus amygdalinus	Madrone and manzanita
Boletus appendiculatus	Oaks and madrone
Boletus calopus	Firs
Boletus dryophilus	Coastal live oak
Boletus edulis	Pines
Boletus flaviporus	Oaks and madrone
Boletus fragrans	Firs
Boletus mirabilis	Hemlock
Boletus regius	Oaks
Boletus satanas	Coastal live oak
Gyroporus castaneus	Oaks
Leccinum arbuticola	Madrone
Leccinum armeniacum	Madrone
Leccinum aurantiacum	Aspen
Leccinum californicum	Aspen
Leccinum constans	Madrone
Leccinum discolor	Aspen
Leccinum insigne	Aspen
Leccinum manzanitae	Madrone and manzanita
Leccinum montanum	Aspen
Pulveroboletus ravenelii	Oaks
Suillus acerbus	Monterey and Bishop pines
Suillus brevipes	Beach and lodgepole pines
Suillus caerulescens	Douglas fir
Suillus fuscotomentosus	Ponderosa pine
Suillus glandulosipes	Bishop pine
Suillus lakei	Douglas fir
Suillus ponderosus	Douglas fir
Suillus pungens	Monterey pine
Suillus tomentosus	Beach and lodgepole pines
Suillus umbonatus	Lodgepole pine
Suillus volcanalis	Jeffrey pine
Tylopilus humilis	Bishop pine
Tylopilus indecisus	Oaks

readily apparent. It is understandable, therefore, that most mycorrhizal associations are only presumed and cannot easily be confirmed except by cultural practices. Such assumptions are based upon repeated observations that a specific bolete always occurs in close proximity to the same kind of tree. In the boletes such associations may be formed with hardwood trees or conifers. Some species appear capable of forming mycorrhizal associations with a rather wide range of trees; others are seemingly highly restricted. In California there are numerous examples of extreme host restriction. *Suillus pungens,* for example, is confined to Monterey and knobcone pines. *Suillus glandulosipes* has been found only under Bishop pine, *S. fuscotomentosus* only with ponderosa pine, and *S. caerulescens, S. lakei,* and *S. ponderosus* only with Douglas fir. Similar associations have been observed with other genera. *Leccinum manzanitae* is known only from under madrone or manzanita and *L. montanum, L. californicum, L. insigne, L. aurantiacum,* and *L. discolor* have been found only under aspen. *Boletus dryophilus, B. aereus,* and *B. regius* have been found only under oaks, and *B. edulis* only under pines. The accompanying table lists several species of boletes from California and their probable mycorrhizal hosts.

THE CALIFORNIA FLORA

The taxonomic and floristic aspects of the bolete flora of California received comparatively little attention prior to 1960. One of the earliest reports is "A Catalogue of the Pacific Coast Fungi" by Harkness and Moore, issued in 1880, in which, as an addendum, sixteen different species of California boletes were listed, mostly from the vicinity of San Rafael in the San Francisco Bay Area. Their list included several European species, which apparently have not been reported since. Unfortunately, most of their collections were destroyed by fire in 1906, and it is strongly suspected that some of the European species recorded by them would now be placed in other taxa. Among those in their list and also in ours are *Boletus piperatus, B. subtomentosus, B. edulis, B. satanas, B. erythropus, Suillus granulatus,* and *Gyroporus castaneus.* Harkness, the senior author, was a physician who lived in the San Francisco Bay Area and pursued his interests in the fungi as a hobby. Most, if not all, of his work was conducted at the California Academy of Sciences in San Francisco. He did an enormous amount of highly professional work not only on the mushrooms of the state but also on other fleshy fungi, particularly the hypogeous species.

Although studies on the fungus flora of California intermittently appeared following the Harkness and Moore report, it was not until after the turn of the century that the next significant papers on boletes were published. In 1905 Earle published descriptions of *Boletus flaviporus* and *B. tomentipes. Boletus flaviporus* is common in the San Francisco Bay Area, but *B. tomentipes* is apparently rare since no recent collections have been made. From 1909 to 1912 Murrill published several papers in which additional species of boletes from the state were included. In 1910, in his treatment of the boletes in *North American*

Flora, he described *Boletus eastwoodiae* as a new species from California. *Suillus lakei* and *Boletus zelleri* were first reported by him from California in 1912. Zeller, from Oregon State University, published frequently on the fleshy fungi of Oregon and occasionally included comments on, or descriptions of, California species.

Following the reports by Murrill, the bolete flora of California was mentioned only incidentally until the appearance of a series of highly significant papers on North American boletes by Snell and Dick and their associates. The first of this series appeared in 1932. Subsequently, many collections of California boletes were sent to them for identification. Some proved to be either new species, including for example, *Suillus megaporinus* and *Boletus atrofuscus,* or species not previously known from North America, such as *Boletus appendiculatus* and *B. regius.* Snell and Dick did not, however, publish any papers devoted exclusively to the California flora.

Alexander H. Smith of the University of Michigan Herbarium spent part of two collecting seasons in the vicinity of Crescent City. He collected numerous boletes as well as other fungi and, although he has not published specifically on his California bolete collections, they are very important. His data and collections were made available. William Bridge Cooke collected fungi at intervals in the Mount Shasta region from 1940–1960, during which time he made the type collection of *Boletus abieticola.* Rolf Singer has also collected in the vicinity of the San Francisco Bay Area, and described *Suillus lithocarpi-sequoiae* from collections made in Muir Woods National Monument.

The total number of bolete species known from California at this time is approximately 80. In comparison, Michigan has 244 (Smith and Thiers); New England, 129 (Snell and Dick); North Carolina, 75 (Coker); Florida, 58 (Singer); Texas and Louisiana, 40 (Thiers); Washington, 35 (Hoffman); Great Britain, 66 (Watling); Central Europe, 65 (Singer). The rather wide numerical discrepancy between these different areas can, in part, be explained by the presence or absence of suitable mycorrhizal hosts. Climate, soil types, and center of origin are other factors that may also play an important role in the distribution patterns and diversity of species. There are in California several cosmopolitan species such as *Boletus edulis, B. piperatus, B. subtomentosus, Leccinum insigne, Suillus granulatus, S. brevipes,* and *Tylopilus pseudoscaber.* As in other bolete floras, *Suillus, Boletus,* and *Leccinum* are the most common genera within the state.

One of the most notable features of the California bolete flora is the absence of species belonging to *Strobilomyces* and *Boletellus.* No taxa belonging to either of these rough-spored genera are presently known west of the Rocky Mountains. Furthermore, there is a relatively small number of species of *Leccinum* and *Boletus* and only a single species of *Gyroporus.* Species of *Fuscoboletinus* and *Boletinellus* have not yet been found. An explanation is not readily available for the absence of the rough-spored species and the reduced number of species of *Boletus* and *Gyroporus.* The scarcity of birches (*Betula* spp.) probably accounts for the reduced number of *Leccinum* species. The absence of species of *Fuscoboletinus* and *Boletinellus* may be similarly explained by the lack of suitable mycorrhizal hosts such as larch (*Larix* spp.). On the other hand, the number of species of *Suillus* found in California is equal to or exceeds that reported for any other bolete flora. This is perhaps due to the great diversity of conifers within the state. The percentage of endemic species of boletes is high, again surpassing that of many other regions. At present there are eleven species of *Boletus,* three of *Gastroboletus,* seven of *Leccinum,* nine of *Suillus,* and two of *Tylopilus* known only from California.

EDIBILITY

Many boletes are well known for their quality as edible fungi. *Boletus edulis* is one of the best of all mushrooms and is excellent when used either fresh or dried. Fortunately for the mycophagist, it is abundant in the state and large collections are made annually in the northern coastal and Sierra Nevada pine forests. Other boletes considered to be top quality are *Boletus zelleri, B. mirabilis, Leccinum aurantiacum, L. manzanitae,* and *L. insigne.* Some species of *Suillus* are also acceptable but generally less popular.

When collecting fleshy fungi for food, care should be taken to avoid including large amounts of soil or substrate. Do not collect older basidiocarps since they are likely to be less palatable and are often insect-riddled. Wrap the collections separately in wax paper to protect them and to avoid possible contamination from other collections. Wax paper is preferable to plastic bags since it permits the air to circulate more freely, thus avoiding sweating or waterlogging. It is not within the scope of this book to recommend cooking methods. There are a number of good recipe books available with directions for preparing mushrooms for the table. Worthy of special mention is the one published by the San Francisco Mycological Society entitled *Kitchen Magic with Mushrooms.*

Most boletes are nonpoisonous but some are regarded as mildly toxic. In this category are those in which the pores are red in contrast to the yellow color of the tubes. *Boletus eastwoodiae,* for example, is known to cause stomach disorders. In some parts of the world there is a feeling against eating any bolete in which the context turns blue when exposed to air; however, some species with this characteristic are not poisonous. It is strongly advised that anyone interested in eating unfamiliar boletes do so *in small amounts* at first so that any unfavorable reactions may be detected.

FIELD AND LABORATORY PROCEDURES

In this study traditional procedures for the handling of specimens were followed. Individual collections were wrapped in wax paper to minimize fading or other color changes due to loss of moisture and to prevent damage to the basidiocarps. In the laboratory suitable specimens were photographed and detailed notes taken on the macroscopic features. Spore prints or masses of basidiospores were obtained from mature caps placed pore surface down on white paper. To maintain high humidity and to avoid air currents while obtaining the prints, the caps were either covered with a glass dish or wrapped with wax paper. Spore prints were thoroughly air dried before the colors were recorded.

The reactions of the various parts of the basidiocarp to different chemical reagents were also determined whenever possible. In the species descriptions the chemical reactions reported for the context and cuticle always refer to the pileus unless otherwise specified. Among the reagents used were the following:

Bases

KOH: 3 percent solution of potassium hydroxide in distilled water.
NH_4OH: 10–14 percent solution of ammonium hydroxide in distilled water.

Acids

HCl: Concentrated solution of hydrochloric acid.
HNO_3: Concentrated solution of nitric acid.
H_2SO_4: Concentrated solution of sulfuric acid.

Miscellaneous

Melzer's reagent, composed of:

Iodine crystals	0.5 grams (g)
Potassium iodide	1.5 g
Distilled water	22.0 g
Chloral hydrate	22.0 g

$FeSO_4$ or $FeCl_3$: 10 percent solution of either ferrous sulfate or ferric chloride in distilled water.
Guaiac: saturated alcoholic solution.

The collections were subsequently dried and placed in small storage boxes until the microscopic data were taken. Microscopic examination of the various parts of the basidiocarp was accomplished by the rehydration of selected portions of the dried basidiocarps and the preparation of freehand sections using a straightedge razor or razor blade. The sections were mounted in 3 percent potassium hydroxide solution (KOH) and in Melzer's reagent. Most microscopic data were taken from the potassium hydroxide preparations. Melzer's reagent was used primarily to detect color changes, including amyloid reactions (dark-blue to blackish discolorations). Color changes were also recorded when observed in the potassium hydroxide mounts. Microscopic measurements were usually made using the oil-immersion objective. Eventually the collections were identified and permanently accessioned into the herbarium.

All the colors recorded in the species descriptions that appear in quotes and parentheses were taken from Ridgeway's *Color Standards and Color Nomenclature.* The type collections as well as all additional collections cited in the species descriptions are deposited in the Cryptogamic Herbarium at San Francisco State University. Duplicate collections have been distributed to several herbaria in the United States, Canada, and Europe.

CLASSIFICATION OF BOLETES

The classification of the boletes "officially" begins with the publication of *Systema Mycologicum* by Elias Fries in 1821. In that publication he placed all species of fleshy pore fungi in the single genus *Boletus.* Soon thereafter various systems for the division of this large taxon into smaller units were proposed, beginning with the appearance, also in 1821, of S. F. Gray's *Natural Arrangement of British Plants.* He placed all the boletes in *Suillus, Leccinum,* or *Pinuzza.* Since the time of these two historic works, other arrangements or systems for the classification of boletes have been presented. Among these was the contribution by Quélet, in which he erected several additional genera. Several years later, Gilbert in *Les Bolets* recognized eleven different genera. Peck was among the first to appreciate the distinctiveness of the bolete flora of the United States, although he placed a majority of the species in previously established European genera. Murrill, on the other hand, recognized few European genera and species as occurring in the United States and, as a consequence, established several new genera native to North America.

Perhaps the most elaborate classification system of the boletes is the one advanced by Singer in 1962. He recognized two families, the Boletaceae and Strobilomycetaceae, and a total of eighteen genera. This system was based on the worldwide bolete flora and has been rather widely adopted, especially in European countries. In 1971, Smith and Thiers, in *The Boletes of Michigan,* followed a system which differed from Singer's in that all species were placed in the family Boletaceae and only eleven genera were recognized. The reduction in the number of genera resulted, in part, from the combining of *Suillus* and *Boletinus,* the elimination of other genera, such as *Xerocomus, Xanthoconium, Porphyrellus,* and *Paragyrodon,* and the transfer of the genus *Phylloporus,* a lamellate genus often classified with the boletes, to the family Paxillaceae, as has been recommended by Watling. Furthermore, because of the obvious affinities of species of *Gastroboletus* to the boletes, this genus was placed in the Boletaceae. The arrangement of genera and species of California boletes as given next is based to a large extent upon the classification system of Smith and Thiers.

Synopsis of California Boletaceae

Genus *BOLETUS*
  Section *Boletus*
    Subsection *Boleti*
      B. aereus; B. edulis; B. mottii
    Subsection *Calopodes*
      B. abieticola; B. appendiculatus; B. calopus var. *frustosus; B. fibrillosus; B. regius*
    Subsection *Luridi*
      B. amygdalinus; B. eastwoodiae; B. erythropus; B. orovillus; B. satanas
  Section *Piperati*
    B. piperatus
  Section *Pseudoboleti*
    B. fragrans; B. pulverulentus; B. tomentipes
  Section *subtomentosi*
    Subsection *Fraterni*
      B. coccyginus
    Subsection *Mirabiles*
      B. mirabilis
    Subsection *Subtomentosi*
      B. chrysenteron; B. dryophilus; B. mendocinensis; B. rubripes; B. smithii; B. spadiceus var. *spadiceus; B. spadiceus* var. *rufobrunneus; B. subtomentosus; B. zelleri*
    Subsection *Versicolores*
      B. flaviporus
  Section *Truncati*
    B. amyloideus; B. truncatus

Genus *GASTROBOLETUS*
  G. amyloideus; G. subalpinus; G. suilloides; G. turbinatus; G. xerocomoides

Genus *GYROPORUS*
  G. castaneus

Genus *LECCINUM*
  Section *Leccinum*
    L. arbuticola; L. armeniacum; L. aurantiacum; L. brunneum; L. constans; L. discolor; L. insigne var. *insigne; L. insigne* var. *brunneum; L. largentii; L. manzanitae* var. *manzanitae; L. manzanitae* var. *angustisporum*
  Section *Scabra*
    L. californicum; L. montanum

Genus *PULVEROBOLETUS*

P. ravenelii

Genus *SUILLUS*

Section *Boletinus*

S. caerulescens; S. imitatus; S. lakei var. *lakei; S. lakei* var. *pseudopictus; S. lithocarpi-sequoiae; S. ponderosus; S. pseudobrevipes*

Section *Suillus*

S. acerbus; S. albidipes; S. borealis; S. brevipes; S. brunnescens; S. californicus; S. fuscotomentosus; S. glandulosipes; S. granulatus; S. luteus; S. megaporinus; S. monticolus; S. punctatipes; S. pungens; S. reticulatus; S. riparius; S. sibiricus; S. subolivaceus; S. tomentosus; S. umbonatus; S. volcanalis

Genus *TYLOPILUS*

Section *Humiles*

T. humilis

Section *Oxydabiles*

T. ammiratii; T. amylosporus; T. ferrugineus; T. indecisus

Section *Porphyrellus*

T. pseudoscaber

TAXONOMIC KEYS AND SPECIES DESCRIPTIONS

BOLETACEAE CHEVALIER
Flore Env. Paris, p. 248. 1926 (as "Ordre")

Homobasidiomycetous; basidiocarp agaricoid, fleshy, putrescent, centrally stipitate; hymenophore tubulose; hymenium on inner surface of hymenophore; basidiospores smooth or variously roughened, globose to ellipsoid to cylindric or terete, brown, yellow to pink or vinaceous in mass; cystidia typically present, often incrusted and sometimes clustered; tube trama bilateral, at least in young basidiocarps; cuticle typically a trichodermium, rarely a cutis; clamp connections present or absent; typically mycorrhizal; terrestrial, rarely lignicolous.

Key to Genera of Boletes in California

1. Hymenophore (tubes) irregularly arranged, not vertically oriented; spore deposit not obtainable — *Gastroboletus*
1. Hymenophore regular in arrangement, more or less vertically oriented; spore deposit typically readily obtained — 2
   2. Tubes and pores with clusters of cystidia in the hymenium that stain dark brown to blackish to occasionally vinaceous in a solution of potassium hydroxide — *Suillus*
   2. Darkly staining, clustered cystidia absent in hymenium — 3
3 Basidiocarp with a dry, pulverulent bright yellow veil (be sure to check young specimens) — *Pulveroboletus*
3. Basidiocarp without a veil — 4
   4. Pores white to pallid when young, becoming yellow at maturity; stipe hollow; spores ellipsoid, yellow in deposit — *Gyroporus*
   4. Basidiocarps without the above combination of characters — 5
5. Surface of stipe with numerous relatively small squamules, which are usually pallid when young but become dark colored (dark brown to black) at maturity — *Leccinum*
5. Surface of stipe not as above — 6
   6. Pores white, yellow, or red; spores brown, olive, olive brown, or yellow brown in deposit — *Boletus*
   6. Pores pink, vinaceous, gray brown to very dark brown (yellow in *T. amylosporus*); spores flesh color to vinaceous brown to reddish brown in deposit — *Tylopilus*

BOLETUS FRIES

Syst. Mycol. I:385. 1821

Pileus small to large or massive, most frequently convex or pulvinate, becoming plane to planoconvex or shallowly depressed; surface frequently dry to moist, rarely subviscid to viscid, glabrous to tomentose to appressed fibrillose or fibrillose scaly, white to yellow to various shades of red, pink, or brown to fuscous or blackish; margin typically entire, with no false veil; context white to pallid or yellow, sometimes with reddish or brownish stains, unchanging or often becoming blue when bruised or exposed, less often changing to other colors; taste mild to bitter or acrid, odor usually mild. Tubes 0.5–2 cm long, white to yellow, often changing color when bruised; pores 0.5–3 mm broad, angular, concolorous with tubes or pink to red to reddish brown, often bluing or changing to some other color when bruised. Stipe equal to clavate to ventricose or bulbous; surface smooth to longitudinally striate to reticulate, dry to moist, rarely viscid, no annulus or veil; color variable, often white to pallid to various shades of yellow, red, or brown. Spores brown to olive brown in mass, smooth, cylindric to subfusoid to subellipsoid; cystidia solitary, often embedded, sometimes staining in KOH or Melzer's reagent, variable in shape and size. Tube trama divergent. Cuticle typically a trichodermium, less commonly a cutis or ixotrichodermium. Clamp connections absent.

Key to Sections

Lead	Section	Go to
1. Some spores in every mount truncate (flattened and/or slightly notched at the apex)	Sect. *Truncati*	
1. Spores not as above		2
2. Tubes and pores reddish to reddish brown; taste of context very acrid	Sect. *Piperati*	
2. Not as above		3
3. Pileus surface velutinous to subtomentose to tomentose or conspicuously fibrillose	Sect. *Subtomentosi*	
3. Pileus surface not as above		4
4. Surface of stipe reticulate, at least in the apical portion	Sect. *Boletus*	
4. Surface of stipe not reticulate		5
5. Pores pink to dark red	Sect. *Boletus,* Subsect. *Luridi*	
5. Pores some shade of yellow, not pink or red	Sect. *Pseudoboleti*	

Boletus Fries

SECTION BOLETUS

Key to Subsections

1. Pores yellow when young	Subsect. *Calopodes*	
1. Pores not yellow when young		2
2. Pores pink to deep red	Subsect. *Luridi*	
2. Pores white when young	Subsect. *Boleti*	

Section Boletus

SUBSECTION BOLETI

Key to Species

1. Pileus strongly reticulate or ridged, brown to dull yellow brown	*B. mottii*	
1. Pileus not as above		2
2. Pileus tan to light brown, surface viscid when wet	*B. edulis*	
2. Pileus dark brown to blackish brown, surface moist to dry	*B. aereus*	

BOLETUS AEREUS *Fries, Syst. Mycol. 1:393. 1821*

Boletus edulis *ssp.* **aerus** (Fries) Konrad and Maublanc, Icon. Sel. Fung. 6:454. 1935.

Illustrations; See Microfiche no. 1

Kawamura, S., Icones of Japanese Fungi, pl. 241.
Leclair, A., and H. Essette, Les Bolets, pl. 35.
Romagnesi, H., Nouvel Atlas des Champignons, pl. 52.
Singer, R., Die Rohrlinge, Teil II, pl. V, figs. 1–6.

Pileus 10–15 cm broad when mature, convex to subglobose when young, becoming broadly convex to plano convex to plane with age; surface moist, never viscid or tacky, often pitted or with broad depressions irregularly distributed, pits and depressions more numerous toward the margin, glabrous but with scattered white pruinose areas or patches; color dark brown to almost black ("warm sepia" to "bister" to "mummy brown" to "chestnut brown"), with age sometimes irregularly fading on the disc to buff or reddish brown ("ochraceous-buff" to "cinnamon-rufous" to "Kaiser brown") with irregular areas almost white to tan ("light buff" to "pale olive-buff"); predominant color, however, much darker; margin slightly incurved and highly irregular when young, becoming decurved and even when older. *Context* 2–4 cm thick, vinaceous when young, with age white, except pale yellow above tubes and pale vinaceous below cuticle. Taste and odor mild.

Tubes 1.5–2.0 cm long, adnate to adnexed, becoming depressed; white when young, becoming yellow ("reed yellow") when mature; *pores* up to 1 mm broad, angular, concolorous, unchanging when bruised.

Stipe 8–11.5 cm long, 2.5–4 cm thick at the apex, clavate to subbulbous when young, unchanging or becoming more or less equal with age, solid; surface dry, reticulate at least over upper half, sometimes strongly and conspicuously so; color when young pallid ("pale pinkish buff" to "pale olive-buff") to almost white, unchanging or becoming vinaceous or vinaceous brown with age. *Context* white, unchanging when exposed.

Spore print dark olive brown. *Spores* 12–13.5 X 4–5 μm, pale ochraceous in KOH and Melzer's, subellipsoid to subfusoid in face view, inequilateral and ventricose in profile, walls thin, smooth.

Basidia 23–30 X 8–10 μm, hyaline in KOH, with ochraceous content in Melzer's, four-spored. *Hymenial cystidia* 28–34 X 7–11 μm, inconspicuous, embedded in the hymenium, hyaline in KOH and Melzer's, thin-walled.

Tube trama subgelatinous in KOH, hyaline to pale ochraceous, subdivergent, hyphae 4–5 μm wide. *Pileus trama* interwoven, homogeneous. *Pileus cuticle* a broad (175 μm) trichodermium of erect, septate hyphal tips, terminal cells not enlarged, not incrusted, hyaline in KOH, pale ochraceous in Melzer's, hyphae 4–5 μm wide. *Clamp connections* absent.

Chemical reactions All tests negative.

Habit, habitat, and distribution Scattered to gregarious under oaks and madrones. It has been found only in the coastal forests of San Mateo County.

Material studied San Mateo County: Thiers 26918, 26926, 26948, 26963, 26978.

Observations The only other fungus known from the west coast with which *B. aereus* might be confused is *B. edulis. Boletus aereus,* however, has a very dark brown to blackish pileus that is moist to dry, but not viscid, shorter spores, and a different type of cuticle.

Singer previously recognized *B. aereus* from the eastern United States, but he did so by reducing such species as *B. variipes* and *B. atkinsonii* to synonymy with it. His action has not been generally accepted, and it appears certain he did not have California material at hand when making his decision. Careful comparison of the California material with European descriptions and collections show that the two are quite similar. Not only is the color of the pilei the same, but convincing similarities are also seen in the nonviscid cuticle (a feature emphasized by Watling), the spore size, and in the microscopic structure of the cuticle (a feature emphasized by Singer). In addition, the basidiocarps were collected under oaks on both continents.

There are two features in which the California material differs from that of Europe. First is the pruinose rather than granulose coating on the pileus and, second, the surface of the stipe and the reticulum are, in general, paler than that usually described for the European species. Neither of these characters seems to be of sufficient magnitude to prevent including them in this species concept.

Edible and delicious.

BOLETUS EDULIS *Fries, Syst. Mycol. 1:392. 1821*

Leccinum Edule (Fries) S. F. Gray, Nat. Arr. Brit. Pls. 1:647. 1821.

Tubiporus edulis (Fries) Karsten, Rev. Mycol. 3:16. 1881.

Dictyopus edulis (Fries) Quélet, Enchir. Fung., p. 159. 1886.

Ceriomyces crassus Murrill, Mycologia 1:149. 1909.

Boletus crasus (Murrill) Jaczewski, Opred. Grib. 1:594. 1913.

Illustrations: See Microfiche No. 8

Kawamura, S., Icones of Japanese Fungi, pl. 239.

Lange, J. E., and M. Lange, 600 Pilze in Farben, p. 191.

Leclair, A., and H. Essette, Les Bolets, pl. 34.

Romagnesi, H., Nouvel Atlas des Champignons, pl. 272.

Singer, R., Die Rohrlinge, Teil II, pl. II, figs. 1–9; pl. III, figs. 1–6; pl. IV, figs. 1–5.

Smith, A. H., The Mushroom Hunter's Field Guide, p. 96, pl. 61.

Smith, A. H., Mushrooms in Their Natural Habitats, Reel V.

Smith, A. H., and H. D. Thiers, The Boletes of Michigan, pls. 14, 147.

Snell, W. H., and E. A. Dick, The Boleti of Northeastern North America, pl. 32.

Pileus 9–18 cm broad when mature, convex to obtusely convex when young, becoming broadly convex to plano-convex at maturity, typically very massive and heavy; surface moist to lubricous, usually becoming viscid to subviscid when wet or with age, uneven to frequently becoming irregularly pitted or wrinkled with age; color when young pale brown to buff ("warm buff" to "ochraceous buff" to "antimony yellow" to "chamois"), with age darkening to brown to reddish brown ("cinnamon brown" to "Mikado brown" to "tawny"), marginal portion often paler than the disc; margin entire, incurved when young, becoming decurved when older. *Context* 2–4 cm thick, firm, white except concolorous with cuticle adjacent to it, unchanging or darkening slightly when exposed. Taste and odor mild.

Tubes 1–3 cm long, adnexed, becoming depressed with age, typically stuffed when young; white when young, becoming yellow ("reed yellow" to "olive-yellow" to "citron green") with age, unchanging when exposed; *pores* very small, two to three per millimeter, angular, concolorous with tubes, unchanging or staining brown when bruised.

Stipe 8–16 cm long, 1.5–2.5 cm thick at the apex, massive, usually conspicuously clavate, up to 6 cm thick at bulbous base, solid, white tomentum at the base; surface dry to moist; white to occasionally pale brown ("cinnamon" to "pinkish cinnamon"), glabrous, reticulate at least over some portion, reticulations sometimes apparent only in apical region with basal portion only uneven or wrinkled. *Context* white, unchanging except for occasional brown stain at the base.

Spore print dark olive brown. *Spores* 13–18 X 4–7 μm, fusiform to subellipsoid, inequilateral in side view, hyaline to pale ochraceous in KOH, dark ochraceous in Melzer's, smooth, walls slightly thickened.

Basidia 30–40 X 10–12 μm, hyaline, clavate, four-spored. *Hymenial cystidia* 48–67 X 5–10 μm, often inconspicuous, hyaline, thin-walled, often deeply embedded in the hymenium, fusoid-ventricose, often narrowed and with tapered, elongated apices.

Tube trama hyaline in KOH, obscurely divergent to subparallel, hyphae 4–7 μm wide. *Pileus trama* homogeneous, loosely interwoven, hyaline, hyphae 5–7 μm wide. *Pileus cuticle* differentiated as a tangled trichodermium, which collapses with age, occasional free hyphal tips; hyphae 5–7 μm wide. *Stipe cuticle* differentiated as a layer of fertile basidia with scattered-to-numerous fusoid-ventricose caulocystidia. *Clamp connections* absent.

Chemical reactions KOH–context pale brown; NH_4OH–context red to dark pink.

Habit, habitat, and distribution Solitary to gregarious in soil in the pine forests over most of California. Fruiting shortly after the onset of the rainy season or after thunderstorms in the mountains. There is little doubt that this species is distributed throughout the state, although no collections have been examined from the southern portion. Generally *B. edulis* is found associated with either coastal or montane species of pines; however, there are some varieties that form mycorrhizal associations with various hardwoods, and this type of habitat should not be overlooked. It is one of our most common species.

Material studied Amador County: Thiers 16924. Butte County: Collett 20; Ripley 1640. Calaveras County: Thiers 16537. Del Norte County: Thiers 17644. Fresno County: Thiers 13372. Humboldt County: Largent 438; Thiers 13984, 14094, 14489, 22887. Lassen County: Thiers 12788. Madera County: Thiers 20906. Marin County: Shervanick 381. Mariposa County: Thiers 20975. Mendocino County: Peters 72; Thiers 8155, 8780, 8851, 9285, 9288, 9340, 10571, 10596, 10646, 11849. Monterey County: Thiers 9416. San Francisco County: Motta 4; Peters 494. San Mateo County: Thiers 11966, 18060. Santa Cruz County: Thiers 10750. Shasta County: Thiers 21588. Sierra County: Thiers 13227. Siskiyou County: Thiers 23492. Sonoma County: Thiers 9414. Trinity County: Thiers 9386. Tuolumne County: Orr 1314; Thiers 12623, 16973, 19638.

Observations This well-known species needs little comment. The basidiocarps are recognized by the yellow-brown to brown, slightly viscid pileus, the white, stuffed tubes, which become yellow with maturity, and the rather massive, often clavate stipe that is reticulated at least at the apex. Numerous subspecies, varieties and forms have been described; however, most of these occur in other countries and are of no concern here. Snell has indicated that ssp. *pinicola* is present in California, and it may be the common form encountered. Until a more thorough and critical study has been done, none of the taxa below the rank of species will be indicated.

In the coastal forests of the northern part of the state *B. fibrillosus* might be confused with *B. edulis*. However, *B. fibrillosus* has a dry, much darker colored, fibrillose pileus, yellow tubes when young, and a much darker stipe. Another species more likely to be confused with *B. edulis* is *B. aereus*, which forms mycorrhizal associations with oak and possibly other hardwoods. *Boletus aereus* is distinguished macroscopically by the very dark color of its glabrous pileus; microscopically, the pileus cuticle is nongelatinous when mounted in KOH. *Boletus mottii* is also similar to *B. edulis*, but is distinguished by a strongly reticulated pileus surface.

The edibility of this bolete is well known to all interested in eating mushrooms for it ranks among the best and is considered excellent in either the fresh or dried state. Unfortunately, fly larvae are often rather common in the inner tissues, which should be carefully checked.

BOLETUS MOTTII *sp. nov.*

Illustrations: See Microfiche No. 14

Pileus 6–12 cm latus, convexus vel planus, humidus vel subviscidus, valde reticulatus vel rugosus, disco brunneus, margine bubalinus. Contextus albus, immutabilis. Sapor et odor mitis. Tubuli 0.5–1 cm longi, albi demum flavi, tactu immutabiles. Stipes 5–7.5 cm longus, 1.5–3 cm crassus, aequalis vel subclavatus, siccus, reticulatus, bubalinus. Sporae 14.3–17.1 X 3.8–5.2 μm, laeves, cylindricae vel subellipsoideae, inaequilaterae. Cystidia nulla. Cuticula intertexta, cellulis terminalibus hypharum ovoideis vel clavatis. Holotypus (Thiers, No. 30502) a Mrs. Lillian Mott lectus prope Grass Valley, Nevada County, November 6, 1972; in Herbarium San Francisco State University conservatus.

Pileus 6–12 cm broad, convex when young, typically becoming plane to plano-convex to shallowly and broadly depressed on the disc at maturity; surface moist to subviscid, conspicuously wrinkled-reticulate over entire surface during all stages of development, the reticulum noticeably raised on the disc, but sometimes becoming shallower and may eventually appear as appressed fibrillose scales on the very margin; the reticulations remain very pronounced on dried pilei and are usually darker than the intervening areas; color brown ("cinnamon" to "cinnamon brown") on the disc, fading to buff ("antimony yellow" to "warm buff" to "light buff" to "light ochraceous buff") toward the margin; margin incurved to decurved, entire. *Context* 1–2 cm thick, soft, white, unchanging when exposed. Taste and odor not distinctive.

Tubes 0.5–1 cm long, depressed, white when young, becoming yellow to greenish yellow ("citrine yellow" to "reed yellow" to "olive yellow") at maturity, unchanging when bruised; *pores* up to 1 mm broad, angular concolorous with the tubes, unchanging when bruised.

Stipe 5–7.5 cm long, 1.5–3 cm thick at the apex, equal to subclavate, solid; surface dry, finely reticulate over entire surface, reticulations suggestive of those of *B. edulis* but more delicate; background color buff to pinkish buff ("pinkish buff" to "pale pinkish cinnamon" to "pinkish cinnamon") during all stages of development; reticulations concolorous with the surface. *Context* white, unchanging upon bruising or exposure.

Spore print color unknown. *Spores* 14.3–17.1 X 3.8–5.2 μm, pale ochraceous in KOH, slightly darker in Melzer's, cylindric to subellipsoid in face view, ventricose and inequilateral in profile, walls smooth, thin; some pleomorphic spores present.

Basidia 23–30 X 7–10 μm, clavate, two- and four-spored, hyaline in KOH. *Hymenial cystidia* not seen (cystidia possibly present, but not reviving in KOH).

Tube trama parallel to only slightly divergent, narrow, laticifers sometimes present, hyphae 3–5 μm wide, cross walls prominent but inamyloid. *Pileus trama* homogeneous, interwoven, hyphae 5–7 μm wide, cross walls inamyloid. *Pileus cuticle* subgelatinous, up to 500 μm broad, well differentiated as a somewhat tangled to suberect trichodermium, hyaline in KOH, not staining in Melzer's, terminal hyphae septate with terminal cells ovoid to clavate, 20–40 X 10–20 μm, subterminal cells more elongate to ellipsoid, walls sometimes incrusted, cross walls inamyloid, hyphae 3–7 μm wide. *Stipe cuticle* differentiated as a layer of fertile basidia with scattered, fusoid to clavate to fusoid-ventricose caulocystidia. *Clamp connections* not seen.

Chemical reactions negative.

Habit, habitat, and distribution Known only from the vicinity of Grass Valley and Nevada City, where it has been found under conifers.

Material studied Nevada County: Thiers 30503 (collection made by Lillian Mott) type.

Observations This species appears most closely related to *B. edulis* and other members of that species complex. The strongly reticulated or ridged surface of the pileus readily distinguishes it, however, from any member of that group. The color of the pileus is also different, and the reticulum on the stipe appears much more delicate than that usually associated with *B. edulis.* The only other species known from California in which the surface of the pileus is even remotely suggestive of *B. mottii* is *B. fibrillosus,* but that species has yellow tubes when young, a rather elongated stipe, and a much darker colored pileus.

Edibility undetermined.

Section Boletus

SUBSECTION CALOPODES

Key to Species

1. Taste of context bitter	*B. calopus* var. *frustosus*	
1. Taste of context mild		2
2. Pileus rose pink to red	*B. regius*	
2. Pileus not colored as above		3
3. Pileus dark brown; surface noticeably fibrillose; context unchanging when exposed	*B. fibrillosus*	
3. Basidiocarp not as above		4
4. Spores 12–14 μm long; hyphae of cuticle not incrusted; pileus glabrous; associated with oaks and other hardwoods	*B. appendiculatus*	
4. Spores 14–17 μm long; hyphae of cuticle with hyaline, spiral incrustations; pileus often with scattered fibrillose scales; associated with firs	*B. abieticola*	

BOLETUS ABIETICOLA *sp. nov.*

Pileus 9–13 cm latus, siccus, appresso-fibrillosus demum fibrilloso-squamosus, subroseus, maculis alutaceis vel flavis. Contextus albus vel subflavus, immutabilis vel raro caerulescens. Sapor et odor mitis. Tubuli 1–2 cm longi, flavi tactu caerulei; pori 1 mm lati, concolores, tactu caerulei. Stipes 9–13 cm longus, 4–5 cm crassus, clavatus vel subbulbosus, siccus, reticulatus, apice flavus, basi rubicosus. Sporae 14–17.5 X 4.5–5.5 μm, ellipsoideae vel subfusoideae vel subcylindricae. Cystidia 40–50 X 4–7 μm, cylindrica vel lageniformia vel subclavata vel subfusoidea. Cuticula intertexta. Holotypus (no. 38653) a Wm. Bridge Cooke lectus prope Mt. Shasta, Siskiyou County, August 8, 1967; in Herbarium San Francisco State University conservatus.

Pileus 9–13 cm broad when mature, convex when young, becoming plano-convex to plane at maturity; surface dry, appressed fibrillose, forming more or less imbricated fibrillose scales that remain readily apparent even when dried, occasionally almost white; color light rose with tan-colored spots interspersed and appearing yellow under the scales; margin incurved, entire. *Context* up to 2 cm thick, white to pale yellow, usually unchanging, occasionally slowly changing to blue in irregular areas when exposed. Taste mild, sweet; odor mild.

Tubes 1–2 cm in length, adnate to adnexed, shallowly depressed with age; yellow ("lemon yellow") when young, darkening and with greenish tints when mature, changing to blue when bruised or exposed; *pores* up to 1 mm broad, angular, concolorous, changing to blue when bruised.

Stipe 9–12 cm long, 4–5 cm thick at the apex, 5–6 cm thick at widest point, clavate to subbulbous, solid; surface dry, strongly reticulate over all except basal portion; color yellow, reddish at base, reticulum yellow; *context* white except reddish at the base, unchanging when exposed.

Spore print color unknown. *Spores* 14–17.5 X 4.5–5.5 μm, pale ochraceous in KOH, ochraceous in Melzer's, ellipsoid to subfusoid to subcylindric in face view, inequilateral and ventricose in profile, walls thin, smooth.

Basidia 22–27 X 8–10 μm, one-, two-, and four-spored, guttulate in KOH, clavate, hyaline; hymenium bright rust-red color in Melzer's, hyaline with scattered yellow areas in KOH. *Hymenial cystidia* 40–50 X 4–7 μm, scattered to rare, inconspicuous, cylindric to pyriform to subclavate to subfusoid, hyaline in KOH and Melzer's, thin-walled.

Tube trama strongly divergent, hyaline and gelatinous in KOH, hyphae 7–8 μm wide. *Pileus trama* interwoven, homogeneous, hyaline in KOH. *Pileus cuticle* a tangled trichodermium of hyphae with occasional erect hyphal tips, hyaline spiral incrustations on most hyphae, contents staining bright red to tawny in Melzer's, hyphae 8–10 μm wide. *Stipe cuticle* composed largely of fertile basidia. *Clamp connections* absent.

Chemical reactions unknown.

Habit, habitat, and distribution Known only from the coniferous forests of higher elevations.

Material studied Siskiyou County: Cooke 38609, 38610, 38611, 38653 type.

Observations This species obviously belongs to the *Boletus appendiculatus–B. fechtneri* complex and has been alternately called one or the other of these for several years. The association with firs (*Abies*) has not been noted for either of the above species. The colors of the pileus, and especially the imbricated, fibrillose scales on the surface of *B. abieticola,* are distinctive; the spores are slightly larger than either *B. appendiculatus* or *B. fechtneri,* and the cystidia are differently shaped. The general arrangement of the spirally incrusted hyphae of the cuticle is also distinctive.

Edibility unknown.

BOLETUS APPENDICULATUS *(Fries) Secretan, Mycogr. Suisse 3:34. 1833*

Boletus edulis f. *appendiculatus* Fries, Syst. Mycol. 1:390. 1821.

Illustrations: See Microfiche No. 4

Leclair, A., and H. Essette, Les Bolets, pl. 39.

Romagnesi, H., Nouvel Atlas des Champignons, pl. 126.

Singer, R., Die Rohrlinge, Teil II, pl. VII, figs. 1–4, 6–9.

Pileus 6–13 cm broad when mature, convex to globose when young, becoming plane to plano-convex to shallowly depressed on the disc, frequently highly irregular and uneven; surface dry to occasionally subviscid in wet weather, dull, glabrous to obscurely tomentose to velutinous, sometimes shallowly rimose to areolate with age; color buff to pale yellow ("ochraceous-buff" to "warm buff" to "light buff" to "antimony yellow") when young, changing to dull yellow to pale brown ("chamois" to "cinnamon-buff") when older, frequently with stains or appearing overlain with rust to red-colored blushes ("ochraceous-tawny" to "russet" to "Corinthian red" to "Eugenia red"), occasionally with dark-brown-colored areas ("Mars brown" to "cinnamon-brown" to "cinnamon-buff"); margin incurved when young, becoming broadly decurved, entire. *Context* 1–3 cm thick, pale yellow to yellow ("cream color" to "naphthalene yellow" to "barium yellow"), unchanging or slowly changing to blue when exposed. Taste and odor not distinctive.

Tubes 1–2 cm long, deeply and broadly to narrowly depressed, when young yellow ("straw yellow" to "barium yellow" to "citron yellow"), becoming dark yellow ("old gold" to "olive lake") with age, bluing intensely when bruised; *pores* up to 1 mm broad, angular, concolorous with tubes, bluing when bruised.

Stipe 6–9 cm long, 2–4 cm thick at the apex, typically bulbous to ventricose, sometimes equal, solid, occasionally with root-like prolongation at base; surface dry, conspicuously to obscurely reticulate for two thirds the entire length, yellow ("barium yellow" to "massicot yellow" to "straw yellow" to "amber yellow"), frequently with vinaceous to reddish brown stains; reticulation concolorous with the surface or sometimes vinaceous to reddish. *Context* yellow in cortical region, whitish in center, unchanging or bluing slightly in irregularly distributed areas when exposed.

Spore print dark olive brown. *Spores* (10)12–14(16) X 4–5(6) μm, appearing amyloid in mass, individual spores appearing brown in Melzer's, pale ochraceous in KOH, smooth, thin-walled, subfusoid to subcylindric.

Basidia 26–31 X 9–12 μm, hyaline, clavate, four-spored. *Hymenial cystidia* 48–65 X 8–13 μm, hyaline, thin-walled, fusoid-ventricose.

Tube trama hyaline, divergent. *Pileus trama* interwoven, homogeneous, amyloid. *Pileus cuticle* differentiated as a trichodermium of interwoven to tangled hyphae, pale brown in KOH, not incrusted. *Stipe cuticle* fertile on the ridges, interwoven in other parts. *Clamp connections* absent.

Chemical reactions KOH—context pale vinaceous; NH_4OH—context blue; HCl—cuticle scarlet; HNO_3—context vinaceous, cuticle scarlet.

Habit, habitat, and distribution Solitary to gregarious in humus in hardwood forests. It is often rather common under oaks in the coastal forests in the Santa Cruz peninsula. With two exceptions, all collections seen by us are from this general area. It quite possibly occurs, perhaps sparingly, throughout the northern coastal forests and in the more mesic portions of the foothills, and should be expected wherever good stands of oaks occur. Like most boletes in this region, basidiocarps appear within two or three weeks after the beginning of the fall rainy season and disappear by the end of December.

Material studied Butte County: Collet 40. Napa County: Thiers 24295. San Mateo County: Sundberg 11-21-62; Thiers 10943, 11967, 14432, 24506. Santa Cruz County: Thiers 10795.

Observations *Boletus appendiculatus* belongs to a rather confusing and difficult group of boletes. There is lack of agreement regarding the concept of this species among the workers in Europe, where it was originally described. Generally, it is characterized by a rather robust basidiocarp with the pileus colored ochraceous to reddish brown, a yellow to almost white context that may or may not change to blue when exposed, yellow tubes that typically change to blue upon bruising, and a yellow to yellow-red stipe that is usually strongly reticulate. The taste is characteristically mild. It is closely related to *B. regius* which was originally described as a variety of *B. appendiculatus. Boletus regius* is easily distinguished, however, by the beautiful old-rose color of the pileus, the often massive size of the basidiocarp, and the rather short, lemon-yellow stipe.

If the identity of this species has been correctly established, it is not only edible but considered choice by some. Because of the general confusion in this taxon, however, it should be eaten with considerable restraint.

BOLETUS CALOPUS *var.* FRUSTOSUS *(Snell and Dick) Miller and Watling, Notes Roy. Botan. Garden, Edinburgh 28:322. 1968*

Boletus frustosus Snell and Dick, Mycologia 33:194. 1941.

Pileus 10–20 cm broad, convex when young, becoming broadly convex to plano-convex at maturity; surface dry, dull, tomentose to fibrillose when young, becoming subtomentose and typically areolate or rimose to frustose with age; color brown ("buffy brown" to "snuff brown") when young, changing to dark yellow brown ("Saccardo's umber") to dark brown ("bister") with age, context as seen in the cracks colored pale tan ("pale olive-buff"); margin typically sterile, entire, incurved. *Context* 2–3 cm thick, pale yellow, quickly changing to blue when exposed, then fading to olive brown ("tawny-olive"). Taste strongly and lingeringly bitter; odor slight or absent.

Tubes 1–2 cm long, adnate when young, becoming deeply and typically narrowly depressed with maturity, pale yellow ("massicot yellow"), becoming darker yellow ("old gold" to "olive lake") with age, changing to blue immediately upon bruising; *pores* 0.5–1.5 mm broad, angular, concolorous with tubes, bluing when bruised.

Stipe 6–14 cm long, 3–6 cm thick at apex, equal to bulbous, solid; surface dry, reticulate, yellow ("barium yellow" to "massicot yellow"), but typically overlain or partially masked with pink to red discolorations; reticulations sometimes yellow over the reddish background, frequently reddish at apex and yellow below. *Context* white to yellow, quickly changing to blue when exposed.

Spore print dark olive brown. *Spores* 14–17 X 4–6 μm, hyaline to pale yellow in KOH, hyaline in Melzer's, ellipsoid to subfusoid in face view, inequilateral and ventricose in side view, smooth, moderately thick walled.

Basidia 30–45 X 10–13 μm, clavate, four-spored, hyaline. *Hymenial cystidia* 33–48 X 5–10 μm, scattered, inconspicuous, hyaline, thin-walled, narrowly fusoid-ventricose to cylindric.

Tube trama hyaline, divergent, with conspicuously amyloid septa. *Pileus trama* interwoven, strongly amyloid, homogeneous, some hyphae with strongly amyloid septa. *Pileus cuticle* differentiated as a loosely interwoven trichodermium staining ochraceous in KOH with numerous free hyphal tips, often showing incrustations, hyphae 6–10 μm wide. *Stipe cuticle* differentiated as a layer of basidia and cystidia similar to the hymenium.

Chemical reaction KOH—context pale vinaceous; NH_4OH—context pale blue; HNO_3—context vinaceous, cuticle scarlet; sulfoformalin—cuticle scarlet; $FeSO_4$—context gray.

Habit, habitat, and distribution Solitary to gregarious in soil under conifers, particularly firs, at higher elevations. Often partially buried in the soil. At present this species is known only from higher elevations in the Sierra Nevada and Cascade ranges; however, it probably occurs in all the mountain ranges where red or white firs are established. *Boletus calopus* is most frequently found in late summer (August–September) following the thunderstorms often prevalent during that time.

Material studied Madera County: Thiers 20875, 20882, 20888, 20913. Mariposa County: Thiers 21052. Mendocino County: Thiers 26904. Nevada County: Thiers 13129, 13156, 13162. Sierra County: Thiers 13204, 13226, 13257, 21186, 23638, 23948, 23949. Siskiyou County: Thiers 23493. Stanislaus County: Thiers 17035. Tuolumne County: Thiers 13274, 17004, 21065, 21113.

Observations This bolete, often massive and heavy, is easily recognized by the dull-brown color of the dry, more or less streaked or fibrillose pileus, which, in dry weather, often becomes severely cracked or areolate, the noticeably bitter taste, and the reticulate stipe, which characteristically shows a combination of red and yellow colors. Microscopically, it is further distinguished by the amyloid reaction of the trama and of the septa or cross walls of the hyphae.

In a recent study Miller and Watling reached the conclusion that typical *B. calopus* does not occur in this country. Their decision was reached after a critical comparison of European and American collections. Whether this decision will be generally adopted is uncertain since similar comparisons have convinced others that the typical form is in this country, particularly in the Pacific Northwest. The variety included here was first collected in the Mount Shasta area by Cooke. The basidiocarps differ from typical *B. calopus* by possessing deep, conspicuous fissures in the pileus and were described as new by Snell and Dick under the name of *B. frustosus.* Miller and Watling, however, decided that these basidiocarps merely represented an extreme areolate or frustose condition of *B. calopus,* and reduced *B. frustosus* to varietal status.

This species is not considered poisonous, but is inedible because of the bitter taste.

BOLETUS FIBRILLOSUS *sp. nov.*

Illustrations: See Microfiche No. 10

Pileus 6–17 cm latus, siccus, tomentosus demum fibrillosus vel fibrilloso-squamosus, brunneus. Contextus albus, immutabilis. Sapor et odor mitis. Tubuli 1.5–2.5 cm longi, pallide flavi, aetate obscure flavi, immutabiles; pori 0.5–1 mm lati, concolores, immutabiles. Stipes 10–16 cm longus, 2–3 cm crassus, siccus, reticulatus, subbrunneus. Sporae 13–17.5 X 3.5–5 µm, subfusoideae vel ventricosae. Cystidia 38–46 X 6–8 µm, subcylindrica vel fusoideo-ventricosa. Cuticula intertexta. Holotypus (no. 8907) a H. D. Thiers lectus prope Jackson State Forest, Mendocino County, November 18, 1961; in Herbarium San Francisco State University conservatus.

Pileus 6–17 cm broad at maturity, convex when young, becoming plano-convex to broadly convex to plane with a flared margin with age; surface dry, even to somewhat wrinkled or rugose, typically tomentose to velutinous, becoming fibrillose, sometimes appearing fibrillose-scaly on the disc; color brown to dark brown ("cinnamon-brown" to "chestnut-brown" to "Mars brown" to "warm sepia" to "burnt umber"), occasionally with paler, buff-colored spots, with age unchanging or becoming somewhat paler on the margin; margin incurved to decurved, becoming plane to flared, entire. *Context* 1.5–3 cm thick, soft, floccose, white to pale buff, darkening slightly near the cuticle, unchanging when exposed. Taste and odor mild.

Tubes 1.5–2.5 cm long, narrowly and shallowly depressed around the stipe, light yellow ("seafoam yellow" to "seafoam green") when young, darkening to dark yellow ("oil yellow" to "olive lake" to "old gold") with age, unchanging or darkening slightly when bruised; *pores* 0.5–1 mm broad, angular, concolorous with tubes, unchanging when bruised.

Stipe 10–16 cm long, 2–3 cm thick at the apex, equal to clavate, rarely tapering, often pinched at the base, solid, white mycelium at the base; surface dry to moist, reticulate over most portions; color pale yellow at apex, otherwise light brown to brown ("cinnamon-buff" to "amber brown" to "hazel"). *Context* white, unchanging when exposed.

Spore print dark olive brown. *Spores* 13–17.5 X 3.5–5.5 µm, pale ochraceous in Melzer's, pale yellow to hyaline in KOH, subfusoid to ventricose in face view, inequilateral in side view, smooth, moderately thickwalled.

Basidia 35–40 X 8–12 µm, hyaline in KOH, clavate, four-spored. *Hymenial cystidia* 38–46 X 6–8 µm, scattered, inconspicuous, hyaline, thin-walled, subcylindric to fusoid-ventricose with elongated apices.

Tube trama hyaline, gelatinous in KOH, divergent. *Pileus trama* interwoven, homogeneous. *Pileus cuticle* differentiated as a trichodermium composed of hyphae having ± equal cells. *Stipe cuticle* differentiated as a layer of fertile basidia and caulocystidia, which are hyaline, subcylindric to clavate, abundant on ridges of reticulum.

Chemical reactions KOH–cuticle red to pink; NH_4OH–cuticle red to pink, HCl–context yellow, cuticle pink to red; HNO_3–context pale yellow, cuticle pink to red; sulfoformalin–cuticle pink to red.

Habit, habitat, and distribution Solitary to scattered in soil in dense mixed coastal forests. It has been most frequently collected in the vicinity of Mendocino, but isolated collections have been made as far north as Crescent City. There are published reports of its presence throughout the Pacific Coast, but it never occurs abundantly. Basidiocarps generally appear early in the fall collecting season, but fruiting continues for only a short time.

Material studied Del Norte County: Thiers 21545. Mendocino County: Breckon 281; Malloch 10-27-62; Motta 9; Sigal 11-14-70; Thiers 8873, 8907 type, 9298, 10581, 10640, 11012, 14608, 18102, 21316, 21526, 24158, 26899, 26936.

Observations This fungus has been commonly referred to as *Boletus olivaceobrunneus,* a species described by Zeller from Oregon. However, Zeller clearly indicated in the type description that *B. olivaceobrunneus* is almost black, has almost black pores, dark-colored spores, and a smooth rather than reticulate stipe. *Boletus fibrillosus* shows none of these characters. Through the courtesy of the New York Botanical Garden, the type of *B. olivaceobrunneus* was examined. It is the opinion of this author that Zeller's type is a collection of *Tylopilus pseudoscaber.* This taxonomic problem is discussed more fully in the discussion of *T. pseudoscaber.*

Boletus fibrillosus is characterized by the dry, dark-brown, fibrillose pileus, pores that are yellow during all stages, a brown, reticulate stipe and a tangled trichodermial type of cuticle. These characters make it obviously distinct from *T. pseudoscaber* and also serve to separate it from *B. edulis,* with which it has been confused in the past. It belongs to the section *Boletus* but does not appear to be very closely related to any known species in that taxon.

Because of the confusion regarding its identity, no definitive statement can be made regarding its edibility. In all probability, however, it is quite safe.

BOLETUS REGIUS *Krombholz, Naturg. Abb. Schw. 2:3. 1832*

Boletus appendiculatus var. *regius* (Krombholz) Konrad, Bull. Soc. Mycol. Fr. 41:63. 1925.

Boletus appendiculatus ssp. *regius* (Krombholz) Quélet, Enchir. Fung. p. 160. 1886.

Illustrations: See Microfiche No. 16

Kawamura, S., Icones of Japanese Fungi, pl. 240.

Leclair, A., and H. Essette, Les Bolets, pl. 40.

Romagnesi, H., Nouvel Atlas des Champignons, pl. 126B.

Singer, R., Die Rohrlinge, Teil II, pl. VII, figs. 8–9; pl. IX, figs. 6–10.

Pileus 8–20 cm broad when mature, convex to broadly convex when young, unchanging or becoming plano-convex to highly irregular in outline with age, typically very massive; surface dry to moist, tomentose to velutinous when young, unchanging or becoming glabrous with age, sometimes highly uneven and conspicuously pitted during all stages of development; color bright pink to rose pink ("Eugenia red" to "jasper red" to "acajou red" to "Pompeian red") during all stages of development, occasionally with brown ("sayal brown") areas when young, and frequently becoming somewhat yellowish toward the margin with age; margin entire, incurved when young, becoming decurved with age. *Context* 1.5–4.5 cm thick, intense yellow ("barium yellow" to "citron yellow" to "picric yellow"), slowly and incompletely changing to blue in spots or irregular areas when exposed. Taste and odor not distinctive.

Tubes 1–2.5 cm long, adnexed when young, becoming deeply and broadly depressed with age, bright yellow ("citron yellow") when young, unchanging or becoming deeper yellow ("empire yellow" to "pinard yellow" to "picric yellow") with age, bluing when exposed; *pores* 0.5–1 mm broad, angular, concolorous with tubes, bluing when bruised.

Stipe 7–11.5 cm long, 3–5 cm thick at the apex, usually conspicuously bulbous, sometimes clavate-bulbous or abruptly bulbous to ventricose, solid; surface dry, finely but distinctly reticulate for two thirds of or entire length, bright yellow ("picric yellow" to "pinard yellow" to "empire yellow"), reticulations concolorous with the surface. *Context* yellow in cortical region, changing to whitish or pallid in center, reddish to pale vinaceous ("avellaneous") at the base, unchanging when exposed.

Spore print pale olive brown to olive brown. *Spores* 12.5–16.5 X 3.7–5 μm, ochraceous in Melzer's, pale ochraceous to hyaline in KOH, ellipsoid to fusoid to subcylindric, smooth, thin-walled.

Basidia 27–32 × 7–9 μm, clavate, hyaline, four-spored. *Hymenial cystidia* 25–45 × 8–17 μm, scattered to numerous, more abundant on the pores, hyaline, thin-walled, clavate to fusoid to subcylindric. *Hymenium* very pale yellow in KOH with occasional yellow amorphous masses interspersed.

Tube trama strongly divergent from a distinct central strand, hyaline. *Pileus trama* interwoven, homogeneous. *Pileus cuticle* differentiated as a trichodermium of tangled to somewhat repent hyphae with free, more or less erect hyphal tips, appearing amyloid. *Hypodermium* appearing gelatinous in KOH. *Stipe cuticle* differentiated as a layer of fertile basidia and hymeniform to vesiculose cells with interspersed caulocystidia similar to the hymenial cystidia. *Clamp connections* absent.

Chemical reactions HNO_3—context pink to pale vinaceous; $FeSO_4$—context gray, cuticle pale lavender.

Habit, habitat, and distribution Solitary to scattered in soil in mixed forests in the coastal regions and under firs at higher elevations. In the coastal forests and other areas at lower elevations, it is usually associated with oaks, but in the absence of this host in the mountains it apparently enters into associations with conifers, especially firs. Fruiting in late August and September in the Sierra Nevada and Cascades; during the fall rainy season along the coast. It is sometimes abundant in the Santa Cruz Peninsula but usually infrequent to rare elsewhere.

Material studied Marin County: Hitsch 12-1-69. Mendocino County: Thiers 9408, 10592, 21418. San Mateo County: Sundberg 391; Thiers 18190, 18337, 26928, 26954, 26979. Santa Cruz County: Malloch 360; Thiers 18552. Shasta County: Thiers 21590. Siskiyou County: Thiers 25382. Sonoma County: Dusatko 10; Thiers 9411. Tuolumne County: Thiers 11226, 12637.

Observations The deep-old-rose to pale-pink pileus, which with age may fade to almost yellow with only pinkish spots or flushes, the yellow context that only erratically changes to blue when exposed, the bright-yellow, noticeably reticulate stipe, and spores 3.5–5.5 μm broad are the distinguishing features of this species. *Boletus regius* is apparently most closely related to *B. speciosus,* but that species has not yet been found in western North America. *Boletus speciosus* has wider spores, a more elongated stipe, and the flesh turns blue immediately when exposed.

In California *B. regius* is most closely related to *B. appendiculatus* and is often confused with it. The ochraceous to rust or rust-brown color of the pileus of *B. appendiculatus* is tne major character separating them. Both species occupy the same type of habitat and, at least in the San Francisco Bay Area, there seems to be considerable intergradation.

Edible, but not considered of high quality.

Section Boletus

SUBSECTION LURIDI

Key to Species

Couplet	Species	Go to
1. Stipe reticulate		2
1. Stipe not reticulate		3
2. Pileus gray to pinkish gray; stipe conspicuously bulbous, pale pink to pallid	*B. satanas*	
2. Pileus brown to reddish brown; stipe clavate to subbulbous, brown	*B. eastwoodiae*	
3. Pileus yellow; spores 5–6 × 3–4 μm	*B. orovillus*	
3. Pileus not yellow; spores larger		4
4. Pileus pink to reddish vinaceous to vinaceous red; pores pink to reddish vinaceous	*B. amygdalinus*	
4. Pileus dark brown to reddish brown; pores red to deep red	*B. erythropus*	

BOLETUS AMYGDALINUS *(Thiers) Thiers, nomen nov.*

Boletus puniceus Thiers, Mycologia 57:529. 1965.

Illustrations: See Microfiche No. 2

Pileus 6–10 cm broad at maturity, convex when young, becoming plano-convex to plane with age, often highly irregular or undulating; surface dry, matted tomentose to cottony fibrillose to appressed fibrillose, usually not appearing fibrillose scaly or rimose; typically colored dark rose to red ("ocher red" to "Etruscan red" to "Corinthian pink") when young, unchanging or changing to dark brownish red ("brownish vinaceous" to "light russet-vinaceous" to "russet" to "vinaceous-russet" to "terra cotta") on the disc when older, sometimes changing to reddish brown ("vinaceous-brown" to "sorghum brown") toward the margin; margin incurved becoming decurved, entire. *Context* 1–2 cm thick, yellow ("naphthalene yellow" to "amber yellow"), bluing immediately upon exposure. Taste and odor not distinctive.

Tubes 1–1.5 cm long, shallowly to deeply depressed, dark yellow ("olive lake" to "pale greenish yellow" to "naphthalene yellow"), bluing upon exposure; *pores* 0.5–1 mm broad, angular, red ("ox-blood red" to "dragons-blood red" to "ferruginous" to "vinaceous-rufous"), bluing immediately upon bruising.

Stipe 5–9 cm long, 1–3 cm thick at the apex, typically bent near the base, equal to ventricose, solid; surface dry to moist, glabrous at the apex, typically becoming tomentose to granulose toward the base, not reticulate; color yellow ("naphthalene yellow" to "massicot yellow"), usually masked with red ("pompeian red" to "dragons-blood red" to "brick red") granules or tomentum, bluing when bruised. *Context* yellow, becoming dark blue immediately upon exposure.

Spore print dark olive brown. *Spores* 11.2–16 X 5.2–8 µm, ochraceous in KOH, dark ochraceous in Melzer's, ellipsoid to subfusoid, thick-walled, smooth, frequently appearing two-celled because of two distinct vacuoles.

Basidia 30–35 X 9–11 µm, clavate, four-spored, guttulate. *Hymenial cystidia* 45–54 X 10–12 µm, scattered, hyaline, fusoid to fusoid-ventricose with tapering apices, relatively obscure, sometimes appearing incrusted.

Tube trama hyaline, obscurely divergent to parallel. *Pileus trama* interwoven, homogeneous. *Pileus cuticle* differentiated as a trichodermium of interwoven hyphae with numerous free hyphal tips, frequently tapering toward the free end, not appearing cystidioid, pale ochraceous in KOH, hyphae ±8 µm wide. *Stipe cuticle* differentiated as a mass of caulocystidia or free hyphal tips, highly irregular in shape and size, ochraceous in KOH, sometimes appearing incrusted. *Clamp connections* absent.

Chemical reactions KOH–context yellow to pale oragne, cuticle red but quickly darkening; NH_4OH–context yellow, cuticle negative or darkening slightly; HCl–context pink; HNO_3 context pink, cuticle red; sulfoformalin–context yellow; guaiac–context pale blue, cuticle pale blue; $FeSO_4$–context negative to pale gray, cuticle negative to pale gray.

Habit, habitat, and distribution Solitary to scattered in humus under madrone and manzanita. It has been found, so far, only in the coastal hardwood forests of central California and in similar forests at lower elevations in the vicinity of Mount Shasta, but it probably occurs throughout the range of madrone and manzanita. It can be easily overlooked because of the similarity in the color of the cap and dead madrone leaves and also because it is frequently completely covered with leaves, never breaking through to the surface. Unlike many boletes, the basidiocarps of this species often do not appear until at least a month after the onset of the fall rains.

Material studied Amador County: Thiers 26915. El Dorado County: Sundberg 1551. Marin County: Hitsch 12-1-69; Thiers 10873. Mendocino County: Thiers 24188. Napa County: Thiers 10821. San Mateo County: Setzer 1529; Thiers 18174, 18341, 26922, 26929, 27020. Santa Cruz County: Thiers 18553. Shasta County: Thiers 21608, 21635.

Observations This species was originally described as *Boletus puniceus;* however, that name had been previously applied to a species from China, thereby necessitating the adoption of a new epithet. *Boletus amygdalinus* shows obvious affinities with *B. erythropus,* but is readily distinguished by the much paler colors of its pileus and tubes. Even old basidiocarps fail to show the dark-brown color of the pileus or the dark-red pores usually associated with *B. erythropus. Boletus amygdalinus* also has larger, more ellipsoid spores, larger cystidia, no laticifers in the pileus trama, and a different type of cuticle on the stipe. The probable mycorrhizal hosts for the two species are also different. Other red-pored boletes occurring in the state are readily distinguished either by the color of the pileus, reticulate stipe, or a combination of these and other characters.

The edibility of this species, so far as known, has not been determined. However, if one abides by the general premise of refraining from eating boletes with red pores, then, obviously, it should be avoided.

BOLETUS EASTWOODIAE *(Murrill) Saccardo and Trotter, Sylloge Fungorum 27:237. 1912*

Suillellus eastwoodiae Murrill, North Am. Flora 9:152. 1910.

Illustrations: See Microfiche No. 7

Smith, A. H., Mushrooms in Their Natural Habitats, Reel V.

Smith, A. H., The Mushroom Hunter's Field Guide, p. 51, pl. 94.

Pileus 10–17 cm broad when mature, convex when young, becoming broadly convex to plano-convex to plane with age; surface dry, typically uneven, pitted to undulating, when young glabrous to obscurely velutinous to tomentose, unchanging or becoming appressed fibrillose with age, old carpophores sometimes appearing rimose-areolate or distinctly fibrillose-scaly; color when young reddish brown ("deep olive-buff" to pale "buffy brown"), frequently with distinct reddish tones or blushes, especially near the margin, typically unchanging with age, or with the tips of the fibrils becoming dark gray; margin entire, incurved when young, becoming decurved with age. *Context* up to 4 cm thick, bright yellow ("picric yellow" to "Naples yellow" to "naphthalene yellow"), quickly changing to dark blue when exposed. Taste and odor not distinctive.

Tubes 0.5–1 cm long, adnate to adnexed when young, typically becoming depressed with age, yellow ("naphthalene yellow" to "barium yellow"), bluing immediately when exposed; *pores* rarely reaching 1 mm in width, angular, when young dark red ("ox-blood red" to "maroon"), when older fading to reddish brown ("garnet brown" to "nopal red" to "Pompeian red" to "dragons-blood red"), sometimes yellowish toward the margin in older pilei, bluing immediately when bruised.

Stipe 8–15 cm long, 2–5 cm thick at the apex, up to 9.5 cm thick at the base, clavate to bulbous (but not abruptly so) to occasionally equal, solid; surface dry, conspicuously reticulate all the way or two thirds the distance to the base; background color pale reddish brown ("apricot buff" to "rufous" to "warm buff"), reticulations red ("Pompeian red" to "dragons-blood red" to "Eugenia red"), base staining brown ("chestnut-brown") when handled, bluing when bruised. *Context* yellow, bluing when exposed.

Spore print brown. *Spores* 13–16 X 5.5–6.5 μm, ochraceous in KOH, dark ochraceous in Melzer's, smooth, moderately thick walled, subellipsoid to subfusoid, ventricose in profile.

Basidia 35–40 X 9–12 μm, one- to four-spored, hyaline in KOH, clavate, hymenium giving a relatively strong amyloid reaction. *Hymenial cystidia* 33–60 X 8–12 μm, scattered, hyaline when young, staining ochraceous in KOH when older, especially those near the pores, fusoid-ventricose to subclavate to basidioid.

Tube trama divergent from a distinct central strand, sometimes appearing gelatinous in KOH. *Pileus trama* interwoven, homogeneous. *Pileus cuticle* differentiated as a trichodermium of tangled unbranched hyphae staining ochraceous in KOH, contents of individual hyphae pale smoky ochraceous, walls noticeably roughened. *Stipe cuticle* differentiated as a compact palisade of globose to pyriform to fusoid-ventricose cells with elongated, tapering apices. *Clamp connections* absent.

Chemical reactions KOH—context pale yellow; HCl—context pink, cuticle pink; HNO_3—context and cuticle pink; sulfoformalin—context and cuticle pink; $FeSO_4$—cuticle gray.

Habit, habitat, and distribution Solitary to gregarious in soil in mixed and coniferous forests. Originally described from the vicinity of San Francisco, it has now been found as far north as Humboldt County. Quite unexpectedly it was once found in abundance under red firs in the vicinity of Yuba Pass at an elevation of approximately 7,000 feet. So far as is known, this is the only report of its presence in localities other than the coastal forest. It must be considered rare, since only one or two basidiocarps per year have been found.

Material studied Humbolt County: Edwards 192. Mendocino County: Thiers 9431, 10591, 10677, 21318, 21421, 21501, 21659, 24490, 26999. Sierra County: Thiers 13218, 13239. Sonoma County: Dusatko November, 1962.

Observations *Boletus eastwoodiae* is distinguished macroscopically most readily by the brown color of the pileus, the red pores, and the stipe, which is frequently clavate or bulbous and strongly reticulate. Microscopically, the size of the spores and the nonamyloid septa in the hyphae of the cuticle and trama contribute toward its distinctiveness. It is quite obviously closely related to *B. satanas,* but that species has smaller spores and amyloid septa in the tramal hyphae. Furthermore, the pileus of *B. satanas* is gray to olive buff with pinkish to vinaceous blushes—not brown, and the stipe typically has an exaggerated bulbous base with the apical portion highly constricted. The color of the pores and stipe are often paler and more vinaceous than those of *B. eastwoodiae.*

The concept of *B. eastwoodiae* is admittedly quite confused. The California collections would be incorrectly determined if the original description was strictly followed, since it clearly states that the stipe is smooth rather than reticulate, and this feature was even used as a key character. Although no report has been found in which the original description was emended, all later descriptions, including those of Murrill, clearly state that the stipe is reticulate. That there is considerable uncertainty regarding the distinction between this species and *B. satanas* is readily apparent when descriptions and photographs of the two are checked.

There is at least one report of poisoning from eating this mushroom. According to this report only "a very small piece of the inner flesh" was swallowed, yet the symptoms were rather severe. Thus, this bolete should not be eaten under any circumstances and care should be taken even in tasting it.

BOLETUS ERYTHROPUS *(Fries) Krombholz, Consp. Fung. Esc., p. 24. 1821.*

Boletus luridus var. *erythropus* Fries, Syst. Mycol. 1:391. 1821.

Illustrations: See Microfiche No. 9

Lange, J. E., and M. Lange, 600 Pilze in Farben, p. 193.
Leclair, A., and H. Essette, Les Bolets, pl. 47.
Romagnesi, H., Nouvel Atlas des Champignons, pl. 130.
Singer, R., Die Rohrlinge, Teil II, pl. XIV, figs. 5–11.

Pileus 8–15 cm broad when mature, convex when young, becoming broadly convex to plano-convex to plane to occasionally umbonate or shallowly depressed with age; surface dry, even, smooth to occasionally rugose to reticulate on the disc, obscurely tomentose to velutinous to occasionally appressed fibrillose when young, frequently glabrous with age; color when young brown ("clay color" to "cinnamon" to "testaceous" to "pecan brown"), with age red ("Prussian red") on the disc and reddish brown ("Hay's russet" to "russet") toward the margin, usually darkening and appearing water-soaked when bruised; margin incurved to broadly decurved, entire. *Context* 1–3 cm thick, firm, compact, yellow ("primrose yellow" to "naphthalene yellow" to "barium yellow"), bluing quickly when first exposed, then often fading to buff ("pale olive-buff"). Taste and odor often pungent, unpleasant.

Tubes 1–2 cm in length, deeply and broadly to narrowly depressed around the stipe, yellow ("pale greenish yellow" to "olive lake" to "primrose yellow"), bluing when bruised; *pores* small, 0.5–1 mm broad, angular, typically red ("Eugenia red" to "jasper red" to "dragons-blood red" to "brick red"), sometimes near reddish orange ("burnt sienna" to "Mars orange"), bluing when bruised.

Stipe 8–12 cm long, 1.5–3.5(6) cm thick at maturity, clavate to subclavate to occasionally equal to tapering toward the apex, solid, with white rhizomorphs attached to the base; surface dry, glabrous to more commonly punctate to obscurely velutinous or fibrillose, not reticulate; typically yellow ("Naples yellow" to "buff-yellow") at the apex, usually with yellow background for entire length but marked with red granules ("Eugenia red" to "acajou red" to "coral red"), basal portion often appearing reddish brown ("Verona brown" to "brick red" to "Hessian brown"). *Context* yellow ("pale greenish yellow" to "Naples yellow") except reddish to reddish vinaceous in the base, bluing when exposed.

Spore print ochraceous brown. *Spores* 13–16 X 4.8–5.5 μm, ochraceous in KOH, dark ochraceous in Melzer's, moderately thick walled, smooth, subcylindric to subfusoid.

Basidia 18–24 X 7–10 μm, hyaline, four-spored, contents granular in KOH. *Hymenial cystidia* 39–45 X 9–11 μm, scattered to numerous, sometimes embedded, subclavate with tapering apices to basidioid, often heavily incrusted and staining bright ochraceous in KOH, occasionally hyaline and apparently not incrusted.

Tube trama obscurely divergent to parallel, hyaline except for irregularly distributed areas colored bright ochraceous in KOH. *Pileus trama* interwoven, with numerous conspicuous, thick-walled laticifers distributed throughout. *Pileus cuticle* a tangled trichodermium composed of moderately thick walled and typically incrusted free hyphal tips. *Stipe cuticle* composed of undifferentiated hyphal tips interspersed with numerous large, inflated, pyriform to subglobose cells staining dark ochraceous in KOH. *Clamp connections* absent.

Chemical reactions KOH—context yellow to pale orange, cuticle red but quickly darkening; NH_4OH—context yellow; HCl context pink; HNO_3—context pink, cuticle red; sulfoformalin—context yellow; guaiac—context pale blue, cuticle pale blue.

Habit, habitat, and distribution Solitary in soil in mixed forests in the coastal regions of central and northern California. A mycorrhizal host has not definitely been established; however, oaks have always been in the vicinity where basidiocarps were found. It is not a common fungus and is apparently highly restricted in its distribution in the state.

Material studied Mendocino County: Cole 11-14-70; Thiers 9519, 10569, 10597, 10625, 14100, 21425, 24161, 24258. Sonoma County: Thiers 9410. Trinity County: Thiers 14268.

Observations The dark-brown pileus, bright-red pores and punctate rather than reticulate stipe surface distinguish this species. It is most likely to be confused with *B. amygdalinus* in California, which, however, is much paler colored, has a more pallid stipe, and has significantly broader spores. The nonreticulated stipe readily distinguishes it from *B. eastwoodiae* and *B. satanas.*

Singer has questioned the presence of this species in North America and has referred collections so named to *B. subvelutipes.* That species, however, is characterized by the presence of a dark red tomentum at the base of the stipe, which is absent in the California specimens.

It is rated as edible by some, but the presence of the red pores should serve as a warning to be very careful when first trying it.

BOLETUS OROVILLUS *Thiers and Kowalski ex Thiers, Mycologia 58:825. 1966*

Pileus 10–15 cm broad at maturity, convex to broadly convex to plano-convex, surface moist, becoming distinctly viscid, subtomentose to velutinous to appressed fibrillose, with age becoming rimose-areolate, particularly on the disc; color during all stages of development a very intense yellow ("picric yellow" to "strontian yellow" to "empire yellow"), slightly darker and duller on the disc with age; margin strongly incurved, becoming decurved, entire. *Context* up to 2 cm thick, yellow, bluing slightly when exposed. Taste and odor mild.

Tubes 0.5–1.5 cm long, adnate, becoming depressed at maturity, yellow, unchanging when exposed; *pores* very small, one to two per millimeter, bright red when fresh and with red exudate, angular, unchanging when bruised.

Stipe 7.5–12.5 cm long, 2–5 cm thick at the apex, ventricose to clavate, sometimes bulbous at the base, yellow mycelium at the base, solid; surface dry, smooth, not reticulate, glabrous to appressed fibrillose to more or less granulose at the apex, more or less concolorous with the pores in apical portion, becoming yellow toward the base. *Context* yellow, changing to pale blue when exposed.

Spore print color unknown. *Spores* 5.5–6.4 X 3–4 μm, pale ochraceous in Melzer's, hyaline to subhyaline in KOH, ellipsoid to subovoid to occasionally narrowed in median portion and somewhat enlarged at the ends, smooth, thin-walled.

Basidia 30–39 X 8–10 μm, hyaline, clavate, four-spored. *Hymenial cystidia* 35–47 X 9–14 μm, scattered to numerous, subclavate to mucronate to somewhat contorted, thin-walled; ochraceous tawny to cinnamon brown in KOH.

Tube trama divergent from a distinct central strand that stains pale ochraceous in KOH. *Pileus trama* amyloid, interwoven, homogeneous, hyaline in KOH. *Pileus cuticle* differentiated as a trichodermium of tangled, more or less interwoven, narrow hyphae, which appear subgelatinous in KOH. *Stipe cuticle* differentiated as a layer of interwoven hyphae. *Clamp connections* absent.

Chemical reactions KOH—cuticle red; HCl—cuticle red at first, then blackening; HNO_3—cuticle red, then blackening; sulfoformalin—cuticle red, then blackening.

Habit, habitat, and distribution Solitary in humus under digger pine (*P. sabiniana* Dougl.). It is known only from the type locality, which is in the vicinity of Oroville. Apparently very rare. Only other known collection is from near Olympia, Washington.

Material studied Butte County: Kowalski 895.

Observations This is one of the most distinctive boletes in California. The intensely yellow pileus, red pores, smooth, nonreticulate stipe, and small spores are the major distinguishing features. It apparently belongs in the subsection *Luridi,* but it is very difficult to determine its relationship among species in that section. The yellow pileus and small spores are features of the subsection *Sulphurei* of the section *Subtomentosi;* however, no species known in that section has red pores and most are lignicolous.

Edibility undetermined, but it should be avoided.

BOLETUS SATANAS *Lenz, Schwamme, p. 67. 1831*

Illustrations: See Microfiche No. 18

Kawamura, S., Icones of Japanese Fungi, pl. 242.
Leclair, A., and H. Essette, Les Bolets, pl. 55.
Romagnesi, H., Nouvel Atlas des Champignons, pl. 131.
Singer, R., Die Rohrlinge, Teil II, pl. XV, figs. 5–10.

Pileus 10–20 cm broad at maturity, convex when young, becoming broadly convex to plano-convex with age, typically massive in appearance; surface dry, unpolished, glabrous to obscurely fibrillose to occasionally fibrillose-scaly, unchanging with age or often becoming diffracted to rimose-areolate on the disc, or becoming obscurely scaly on the disc and finely velutinous to subtomentose to glabrous toward the margin; color pale gray ("pale olive-buff" to "buffy brown") with blushes colored pink ("alizarine pink" to "old rose" to "thulite pink" to "pale rosolane purple" to "spinel pink"); young basidiocarps often predominantly gray ("pale olive-buff") with only faint traces of the pink coloration, but in older pilei the pink shades typically predominate, particularly near the margin; in some very old pilei the pink pigment is completely absent and the pileus colored near olive gray ("citrine-drab" to "dark olive-buff"); margin entire to eroded, incurved, becoming decurved with age. *Context* 2–5 cm thick, yellow ("pale olive-buff" to near "glass green" to "deep seafoam green"), bluing readily on exposure when young, sometimes erratically so when older. Taste and odor not distinctive.

Tubes 1–2 cm in length, arcuate-decurrent to subdecurrent when young, becoming deeply to shallowly depressed with age, yellow ("pale greenish yellow" to "deep seafoam green" to "pale glass green"), changing to blue when exposed or bruised; *pores* very small, up to two per millimeter, when young reddish brown ("madder brown" to "garnet brown" to "ox-blood red" to "Pompeian red"), when older often fading to red ("Eugenia red" to "jasper red" to "dragons-blood red"), bluing when bruised.

Stipe 6–12 cm long, 3.5–7 cm thick at the apex, conspicuously and abruptly bulbous, bulb 9–14 cm broad, solid; surface dry, conspicuously reticulate over upper portion or entire surface, glabrous to obscurely tomentose over nonreticulate portion, sometimes streaked and appearing appressed-fibrillose; color similar to the pileus, the reticulations colored pink to vinaceous, bulb often pinkish vinaceous ("vinaceous-pink" to "alizarine pink"). *Context* concolorous with the pileus, bluing when exposed, except in basal portion.

Spore print brown. *Spores* typically 11–14.4 × 4–6 μm, occasional giant spore up to 17.6 X 8 or 16 X 4 μm, moderately thick walled, smooth, subellipsoid to subfusoid to subcylindric, ochraceous in KOH, bright ochraceous in Melzer's.

Basidia 24–28 X 7–10 μm, ochraceous in KOH, four-spored, clavate. *Hymenial cystidia* 28–42 X 7–11 μm, deeply embedded, often obscure, hyaline when young often becoming ochraceous with age, clavate to fusoid to basidioid.

Tube trama obscurely divergent from central strand, hyaline to yellow in KOH; hyphal walls typically appearing incrusted or nodulose. *Pileus trama* interwoven, homogeneous, hyaline in KOH, some septa amyloid, hyphae 3–5 μm in diameter. *Pileus cuticle* differentiated as a loosely interwoven trichodermium, hyaline to ochraceous in KOH. *Stipe cuticle* differentiated as a closely packed turf of inflated to cylindric hyphal tips. *Clamp connections* absent.

Chemical reactions KOH–context pale yellow, cuticle brownish gray; HCl–context pale yellow; HNO_3–cuticle pink; sulfoformalin–context yellow, cuticle pink; $FeSO_4$–cuticle brownish gray; guaiac–context blue.

Habit, habitat, and distribution Solitary to gregarious in humus under oaks. At present this species is known only in the coastal forests. It appears to be most abundant in the Santa Cruz Peninsula, where it is usually found associated with coastal live oaks. *Boletus satanas* is not common, but may be locally abundant during some seasons.

Material studied Marin County: Hitsch 12-1-69. Mendocino County: Thiers 9295, 21321, 26891. San Mateo County: Setzer 1404; Thieme 11-15-69; Thiers 9445, 26923, 26950, 26975. Shasta County: Thiers 21585, 21677. Sierra County: Thiers 13218, 13239. Sonoma County: Dusatko 3, 9. Tuolumne County: Thiers 21108.

Observations The California collections of this species match the descriptions and illustrations of European material quite satisfactorily, although some foreign workers appear somewhat reticent to admit its presence in North America. It is easily recognized by its massive size, the pale-pinkish to vinaceous flushes in an otherwise light-gray to olive-buff-colored pileus, and, perhaps most readily, by the immensely bulbous base of the stipe with the apex greatly constricted. The pores are characteristically pink to vinaceous rather than darker red. As indicated previously, it is most likely to be confused with *B. eastwoodiae.* That species, however, has a brown pileus, dark-red pores, and the stipe is often only clavate. Microscopically, the spores of *B. satanas* are smaller and the septa of some of the tramal hyphae are amyloid. For further discussion, see comments under *B. eastwoodiae.*

The presence of pink or reddish pores places this species in the group of boletes generally not recommended for the table. It should be noted that there are families in the San Francisco Bay Area who claim to have eaten it over a long period of time and to have never suffered any ill effects. However, these basidiocarps were eaten only after they were thoroughly cooked. Remember, if you decide to try it, ingest only very small amounts at first until your reaction to it becomes apparent.

SECTION PIPERATI

Key to Species

Only one species known from California	*B. piperatus*

BOLETUS PIPERATUS *Fries, Syst. Mycol. 1:388. 1821*

Suillus piperatus (Fries) O. Kuntze, Rev. Gen. Pl. 3:535. 1898.
Ceriomyces piperatus (Fries) Murrill, Mycologia 1:150. 1909.
Ceriomyces ferruginatus (Fries) Murrill, North Am. Flora 9:143. 1910.

Illustrations: See Microfiche No. 15

Lange, J. E., and M. Lange, 600 Pilze in Farben, p. 189.
Leclair, A., and H. Essette, Les Bolets, pl. 18.
Romagnesi, H., Nouvel Atlas des Champignons, pl. 121.
Singer, R., Die Rohrlinge, Teil I, pl. X, figs. 6–10.
Smith, A. H., and H. D. Thiers, The Boletes of Michigan, pls. 114–115.
Snell, W. H., and E. A. Dick, The Boleti of Northeastern North America, pl. 23.

Pileus 3–7 cm broad, convex to broadly convex when young, becoming plano-convex to plane or occasionally subumbonate with age; surface moist to dry, often subviscid with age, typically becoming subviscid when wet, dull, glabrous to obscurely appressed fibrillose, sometimes old basidiocarps appearing obscurely rimose or slightly fibrillose-scaly; color reddish brown to brown ("pinkish cinnamon" to "russet" to "mahogany red") when young, with age fading to pale reddish brown ("cinnamon-rufous" to "Hay's russet" to "vinaceous rufous"); margin incurved to decurved, becoming decurved to plane or flared, entire. *Context* 0.5–1.5 cm thick, yellow ("straw-yellow" to "warm buff") changing to pale vinaceous to pink ("avellaneous") near the cuticle, unchanging or darkly pigmented areas changing to grayish vinaceous when exposed, not changing to blue. Taste strongly acrid; odor not distinctive.

Tubes 0.5–1 cm long, adnate to arcuate-decurrent when young, becoming depressed around the stipe with age, yellow to reddish yellow ("honey yellow" to "Isabella color" to "hazel" to "Hay's russet" to "ferruginous"), unchanging or darkening slightly when bruised; *pores* relatively large, 1–2 mm broad, angular, red to reddish brown.

Stipe 2–4 cm long, 0.5–1 cm thick at the apex, equal or tapering toward the base, solid, with copious bright-yellow mycelium at the base; surface dry, glabrous to occasionally slightly appressed-fibrillose, reddish brown ("Hay's russet"), usually more or less concolorous with the surface of the pileus. *Context* yellow, unchanging or darkening slightly when exposed.

Spore print cinnamon brown. *Spores* 8.5–12 X 3–4 µm, rust brown in Melzer's, hyaline to pale ochraceous in KOH, subfusoid to subellipsoid in face view, slightly inequilateral in side view.

Basidia clavate, four-spored, hyaline in KOH, 24–29 X 8–10 µm, hymenium dark ochraceous in Melzer's. *Hymenial cystidia* 45–70 X 10–15 µm, abundant, conspicuous, hyaline, fusoid-ventricose to subclavate to fusoid with obtuse apices.

Tube trama hyaline, gelatinous, divergent, hyphae 4–6 µm wide. *Pileus trama* loosely interwoven, homogeneous, hyphae 4–6 µm wide. *Pileus cuticle* differentiated as a tangled trichodermium staining dark brown in KOH and rust brown in Melzer's. *Stipe cuticle* differentiated as a layer of fertile basidia and caulocystidia similar to those in the hymenium, with occasional areas in which basidia and cystidia are lacking and surface appears as a layer of interwoven hyphae. *Clamp connections* absent.

Chemical reactions NH_4OH–context dark vinaceous; KOH–context dark reddish brown; HCl–cuticle pink; $FeSO_4$–context grayish.

Habit, habitat, and distribution Solitary to scattered in humus in the mixed coastal forests of northern California. It appears to be quite common from Mendocino County northward, but is apparently rare from the Santa Cruz Peninsula southward.

Material studied Del Norte County: Ripley 865; Thiers 17720, 21353. Humboldt County: Jonas 157; Sundberg 895; Thiers 14082, 14281, 14456, 14523, 17641, 17743, 18254, 21358, 21499, 22914, 23021, 26913. Mendocino County: Breckon 269; Jensen 33; Motta 238; Thiers 8753, 8792, 14102, 14109, 14148, 17867, 24222. San Francisco County: Thiers 24710, 24733. San Mateo County: Thiers 27018. Sonoma County: Breckon 288.

Observations This is probably the smallest species of *Boletus* from California with the cap often not more than two or three centimeters in diameter. It is further distinguished by the reddish to rust-brown color of the pores, the bright-yellow color of the context in the base of the stipe, the bright-yellow mycelium attached to the basal part of the stipe, and the very acrid taste. There is no other bolete known from California with a similar taste. The context of the pileus does not change to blue upon exposure, but this feature should always be checked since *B. piperatoides*, recently described from Michigan, is similar in most other characters except for the bluing of the context.

This species has been placed in the genus *Suillus* by Singer and Snell; however, it does not seem to have enough similar characters to warrant such action.

According to McIlvaine, the peppery taste is lost upon cooking and it is then edible and delicious. On the other hand, Murrill warns that it is poisonous and should be avoided.

SECTION PSEUDOBOLETI

Key to Species

1. Tubes changing to brick red when bruised or dried	*B. tomentipes*	
1. Tubes changing to blue when bruised		2
2. Pileus up to 10 cm broad, brown on the margin; hyphae of cuticle not incrusted; stipe cuticle differentiated as a cutis	*B. pulverulentus*	
2. Pileus up to 15 cm broad, tan to buff on the margin; hyphae of cuticle incrusted; stipe cuticle differentiated as a layer of basidia and caulocystidia	*B. fragrans*	

BOLETUS FRAGRANS *Vittadini, Funghi Mang., p. 153. 1835*

Xerocomus fragrans (Vittadini) Konrad and Maublanc, Encycl. Mycol. 20:120. 1952.

Illustrations: See Microfiche No. 12

Leclair, A., and H. Essette, Les Bolets, pl. 38.

Singer, R., Die Rohrlinge, Teil II, pl. XII, figs. 1–7.

Pileus 6–15 cm broad at maturity, convex, becoming broadly convex to plano-convex to highly irregular in outline; surface dry, velutinous to finely tomentose and sometimes with a band of pallid, appressed fibrillose scales toward the margin when young, usually becoming glabrous to subglabrous with age, generally areolate to rimose-areolate when older; color olive brown to brown ("cinnamon" to "buffy brown" to "wood brown") on the disc, fading to buff ("olive-buff" to "pale olive-buff" to "warm buff") toward the margin; margin entire, incurved. *Context* 1–3 cm thick, yellow ("primrose yellow" to "cream color") changing to blue when exposed; taste and odor mild.

Tubes 0.5–1 cm in length, when young bright yellow ("primuline yellow") changing to dark yellow ("colonial buff" to "old gold") when older, bluing when bruised; *pores* up to 1 mm broad, angular, concolorous with tubes, bluing when bruised.

Stipe 4–10 cm long, 1–3.5 cm thick at apex, equal to subclavate to sometimes tapering slightly at the base, solid; surface dry, glabrous to obscurely furfuraceous to punctate, not reticulate; color yellow ("primrose yellow") at the apex, changing to tan ("warm buff" to "buffy brown") toward the base, occasionally with a reddish band at the apex. *Context* yellow to concolorous with the surface, bluing when bruised.

Spore print dark olive brown. *Spores* 13–17.5 × 5–7.8 μm, subfusoid, to subcylindric in face view, ventricose in side view, ochraceous in KOH and Melzer's, smooth, thin-walled.

Basidia 28–35 × 6–9 μm, hyaline in KOH, clavate, four-spored. *Hymenial cystidia* apparently not differentiated.

Tube trama strongly divergent from a narrow central strand, hyaline in KOH, dark ochraceous in Melzer's, hyphae 4–6 μm wide, often with scattered laticiferous hyphae. *Pileus trama* homogeneous, interwoven. *Pileus cuticle* differentiated as a trichodermium of loosely interwoven hyphae with occasional free hyphal tips, hyphae 5–7 μm wide, with abundant hyaline, spiral incrustations that are ochraceous in KOH, bright ochraceous in Melzer's. *Stipe cuticle* differentiated as a layer of fertile basidia with scattered caulocystidia 27–35 × 8–12 μm, which are hyaline, thin-walled, fusoid to fusoid-ventricose in outline. *Clamp connections* absent.

Chemical reactions unknown.

Habit, habitat, and distribution Scattered to solitary in soil under conifers. This species is very uncommon in California and has been found only in the Yuba Pass and Luther Pass areas.

Material studied El Dorado County: Thiers 32756. Sierra County: Thiers 21142, 21181, 23637.

Observations This is one of the large, massive boletes that occurs, as far as is known, only in the alpine areas, where it appears to be associated with red fir. It is distinguished by the dull, brown pileus, pallid flesh that stains blue, and a nonreticulate stipe. It is most likely to be confused with *B. calopus,* but can be easily distinguished from it by the smooth rather than reticulate stipe.

The California species differs from the European species in the apparent mycorrhizal associate. Almost invariably it is reported as associated with oaks in Europe, whereas in California it is only found under conifers. Since no other significant difference has been observed, the present disposition of this species seems to be the most satisfactory.

The European species is edible according to Bresadola; however, the edibility has not been determined for the California specimens.

BOLETUS PULVERULENTUS *Opatowski, Wiegm. Archiv. Naturgesch. 2:27. 1836*

Xerocomus pulverulentus (Opatowski) Gilbert, Les Bolets, p. 116. 1931.

Boleıus mutabilis Morgan, J. Cincinnati Soc. Nat. His. 7:6. 1884.

Ceriomyces cyaneitinctus Murrill, Lloydia 6:225. 1943.

Boletus cyaneitinctus (Murrill) Murrill, Lloydia 6:228. 1943.

Illustrations

Lange, J. E., and M. Lange, 600 Pilze in Farben, p. 187.

Leclair, A., and H. Essette, Les Bolets, pl. 26.

Singer, R., Die Rohrlinge, Teil II, pl. X, figs. 13–18.

Smith, A. H., and H. D. Thiers, The Boletes of Michigan, pls. 122, 123.

Snell, W. H., and E. A. Dick, The Boleti of Northeastern North America, pl. 42.

Pileus 5–8 cm broad at maturity, broadly convex when young, becoming plane with slightly elevated or uplifted margin when older; surface dry, dull, velutinous to subtomentose, typically becoming noticeably areolate with age; color brown to dark brown ("Prout's brown" to "cinnamon-brown") to sometimes as pale as buff ("warm buff"), occasionally with rust-colored tinges or flushes when older; margin entire, decurved, becoming plane or uplifted. *Context* up to 1 cm thick, yellow ("Martius yellow"), changing to blue immediately upon exposure. Taste and odor mild.

Tubes 1 cm in length, decurrent to adnate, greenish yellow ("light greenish yellow") when young, becoming olive yellow ("olive lake" to "sulphin yellow") when older, bluing immediately when bruised; *pores* 1 mm broad, angular, yellow, bluing quickly when bruised.

Stipe 4–7 cm long, 1–2.5 cm thick at the apex, equal or tapering toward the base, stuffed; surface dry, obscurely punctate or furfuraceous, sometimes longitudinally ridged, not reticulate, bright yellow ("Martius yellow") at the apex, changing to deep yellow ("antimony yellow") in midportion and typically reddish brown ("russet" to "Mars brown") toward the base, bluing when handled. *Context* yellow, turning blue immediately upon exposure.

Spore print dark olive to olive brown. *Spores* 14.0–17.5 × 4–6.5 μm, ochraceous in KOH, weakly amyloid, subellipsoid to subfusoid in face view, ventricose in side view, many appearing obscurely truncate with no annular thickening and no germ pore, smooth, moderately thick walled.

Basidia 25–32 X 6–9 μm, four-spored, clavate, hyaline. *Hymenial cystidia* 45–60 X 10–16 μm, scattered to rare, hyaline, thin-walled, fusoid to clavate, often with elongated tapering apices.

Tube trama obscurely divergent from mediostratum, which stains brown in KOH, hyphae 3–5 μm wide. *Pileus trama* homogeneous, interwoven. *Pileus cuticle* differentiated as a tangled trichodermium staining dark brown in KOH, with numerous free, erect hyphal tips. *Stipe cuticle* not differentiated. *Clamp connections* absent.

Chemical reactions According to Singer, the following reactions occur: KOH–cuticle darkening; HNO_3–cuticle red.

Habit, habitat, and distribution Solitary in humus in mixed coastal forest. Known only from a single collection made in the vicinity of Eureka. It is apparently more abundant in Washington, Oregon, and Idaho, where several collections have been made.

Material studied Trinity County: Lamphere 12-6-62.

Observations The inclusion of this species is based upon a collection consisting of only a single fruit body. Had it not been for an abundance of material from the adjacent northwestern states, it would have been omitted. It is distinguished by the dull, dark-brown, tomentose pileus, which in the western United States characteristically becomes rimose-areolate with age, and by the immediate change to blue of all parts of the basidiocarp when bruised or exposed. As indicated, some spores are frequently obscurely truncate, and it is interesting to speculate as to the significance of this feature from a phylogenetic point of view. In view of the recent discovery of at least three species having distinctly truncate spores, it is quite possible that *B. pulverulentus* represents an intermediate stage.

Edible, but so rare as to be of no significance.

BOLETUS TOMENTIPES *Earle, Bull. N.Y. Botan. Garden 3:298. 1904*

Ceriomyces tomentipes Murrill, Mycologia 1:154. 1909.

"Pileus thick, convex to expanded, 9–13 cm broad, about 3 cm thick; surface dry, minutely tomentose to glabrous, umbrinous; context whitish or discolored, changing to blue when wounded; hymenium ventricose, deeply and broadly sinuate-depressed, decurrent; tubes sordid-yellow, becoming brick-red when wounded or on drying, mouths small, about 1 mm broad, circular; spores ellipsoid, brownish, about 14 X 7 μ [μm]; stipe cylindric, densely but minutely velvety-pubescent, sometimes becoming nearly glabrous above, brick-red, flecked brown below, solid, 8–13 cm long, 2.5–3.5 cm thick.

"Found only at Stanford University, California, among decaying oak leaves."

The above is a copy of the original description. As far as can be determined, no recent collections of this bolete have been made. Through the courtesy of Dr. Clark Rogerson of the New York Botanical Garden, the type and one additional collection have been examined.

There are two packets labeled "Type." The label on one, handwritten, has No. 132, "collected by C. F. Baker, Nov. 30, 1901." The second packet has a more recent label of the New York Botanical Garden, is stamped "Type," and is labeled *Ceriomyces tomentipes.* A third collection, not labeled "Type" and with a label headed "Pacific Slope Fungi," distributed by C. F. Baker, has the No. 132 and is presumably part of the type collection. The label on this last collection has the following additional information on it: "Collected in foothills near Stanford University, Santa Clara County, California, Nov. 30, 1901. Coll. C. F. Baker." "Frequent among decaying leaves under oak trees." "Well marked by its velvety, opaque brown pileus, the yellowish to brick red pores and the flesh changing to fine blue." The following additional notes are included in the packet:

"Habitat: decaying leaves.
"Size: 9–13 cm in diameter, 3 cm thick.
"Color of pileus—clear brown, opaque.
"Surface with a glaucous, almost velvety appearance.
"Margin entire.
"Pore attachment—deeply, suddenly and broadly sinuate, then slightly decurrent.
"Shape of pores—small.
"Color of pores—sordid yellowish becoming brick red when bruised and in drying.

"Spores–brownish, 14 X 7 μ [μm].
"Veil none. Annulus none.
"Stipe 8–13 cm long, 2.5–3.5 cm in diameter.
"Shape: cylindrical.
"Surface smooth.
"Color brick red, finely speckled with brown below, solid.
"No volva.
"Flesh white to brownish, changes to a fine blue when bruised.
"Taste and smell–as a normal agaric.
"A magnificent boletus."

The collections, generally, appear as if they were poorly dried, are some shade of brown, and show none of the colors indicated for fresh basidiocarps.

A microscopic examination of the type was made and yielded the following additional data: *spores* ochraceous in KOH, 12–14 X 5–6 μm, fusoid to subellipsoid in face view, inequilateral in profile, thin-walled, not dextrinoid or amyloid. *Cystidia* 23–25 X 7–9 μm, fusoid to fusoid-ventricose, hyaline, thin-walled. *Tube trama* divergent, hyaline. *Pileus trama* interwoven, homogeneous. *Pileus cuticle* differentiated as a trichodermium of tangled to repent hyphae. *Clamp connections* absent.

Although no recent collections of *B. tomentipes* are known, its description is included here in the hope that this interesting fungus might be rediscovered. The brick-red discoloration of the tubes when bruised or dried is a most distinctive feature.

Edibility unknown.

SECTION SUBTOMENTOSI

Key to Subsections

1. Spores up to 22 μm in length; pores yellow	Subsect. *Mirabiles*	
1. Spores less than 20 μm in length, or if greater than that in length, then pores pink or reddish		2
2. Pileus red or reddish vinaceous	Subsect. *Fraterni*	
2. Pileus not colored as above		3
3. Tubes and pores staining blue when bruised	Subsect. *Subtomentosi*	
3. Tubes and pores not changing to blue when bruised	Subsect. *Versicolores*	

Boletus Fries

Section Subtomentosi
SUBSECTION FRATERNI
Key to Species

Only one species known from California *B. coccyginus*

BOLETUS COCCYGINUS *sp. nov.*

Pileus 2–4 cm latus, siccus, velutinus demum glaber, coccyginus. Contextus flavus tactu immutabilis vel tarde caeruleus. Sapor et odor mitis. Tubuli 2–5 mm longi, sordide flavi, immutabiles; pori concolores, immutabiles. Stipes 1.5–4.5 cm longus, 0.5–1 cm crassus, siccus, appresso-fibrillosus demum glaber, laevis, rufus; contextus flavus, immutabilis. Sporae 11.2–17.6 X 5.3–7 µm, amyloideae, cylindricae, laeves. Cystidia absentia. Cuticula intertexta. Holotypus (no. 21237) a Wm. Jordan lectus prope Big Meadow Campground, Alpine County, October 14, 1967; in Herbarium San Francisco State University conservatus.

Pileus 2–4 cm broad at maturity, convex when young, becoming broadly convex to plane to irregularly lobed, incised or convoluted at maturity; surface dry to moist but not viscid, typically finely velutinous to covered with a fine red pubescence when young, usually becoming glabrous with age, plush-like in appearance, sometimes finely areolate near the margin, more or less evenly colored purple red ("Corinthian red" to "dark vinaceous" to "deep Corinthian red"), sometimes slightly paler ("orange-vinaceous" to "Etruscan red"), unchanging or becoming slightly paler with age; margin entire, decurved. *Context* 0.5–1 cm thick, yellow ("maize yellow" to "baryta yellow"), usually unchanging, but sometimes slowly and erratically changing to blue. Taste mild; odor not distinctive.

Tubes 2–5 mm long, adnate to depressed, seceding from stipe, dull yellow ("old gold" to "colonial buff"), unchanging when exposed; *pores* 0.3–1 mm broad, angular, bright yellow ("amber yellow"), unchanging when bruised.

Stipe 1.5–4.5 cm long, 0.5–1 cm thick at apex, equal to tapering downward, light-yellow mycelium at base, stuffed to solid; surface dry, appressed fibrillose becoming glabrous, smooth or with longitudinal striations, not reticulate; color red ("Corinthian red" to "Etruscan red"), yellowish at the base. *Context* yellow, unchanging when exposed.

Spore print olive brown. *Spores* 11.2–17.6 X 5.3–7 µm, ochraceous in KOH, often dextrinoid, cylindric, to ovoid to ellipsoid, highly variable in size and shape, pleomorphic, smooth, moderately thick walled.

Basidia 23–28 X 7–10 µm, apparently one-, two-, three-, and four-spored, clavate, hyaline in KOH. *Hymenial cystidia* not apparent, but occasional basidioid cells staining dark brown in Melzer's scattered in hymenium.

Tube trama divergent, hyaline in KOH, hyphae up to 6 µm wide. *Pileus trama* closely interwoven, pale ochraceous in KOH, homogeneous. *Pileus cuticle* differentiated as a tangled trichodermium of more or less repent hyphae 5–7 µm wide, staining dark yellow in KOH, ochraceous in Melzer's, walls asperulate to faintly incrusted. *Clamp connections* absent.

Chemical reactions unknown.

Habit, habitat, and distribution Gregarious to subcespitose in soil in conifer forests. This fungus is apparently rare and is known from only three collections from California plus a few from the Pacific Northwest. In California it is curious to note that, even though known from only three collections, it has been found in the coastal and mountainous regions.

Material studied Calaveras County: Thiers 21237 (collection made by William Jordan) type. San Francisco County: Orr 919. San Mateo County: Orr 860.

Observations *Boletus coccyginus* apparently belongs in the section *Subtomentosi,* but does not fit well into any subsection recognized within that section. Except for the fact that the pores do not stain blue when bruised, it would seem to belong in the subsection *Fraterni,* where it is tentatively placed. It appears most closely related to *B. campestris,* which has pores that turn blue when bruised, noticeable cystidia, and a pileus that is ferrugineous rather than rose red when dried. *Boletus campestris* occurs in a different type of habitat and is not known from California.

Boletus coccyginus is very distinct and is not likely to be confused with any other California bolete. It is characterized by the rose-colored pileus, small pores that do not change to blue when bruised, no highly differentiated cystidia, and pleomorphic (variable) spores.

Edibility undetermined, but its small size makes it insignificant for culinary purposes.

Section Subtomentosi

SUBSECTION MIRABILES

Key to Species

Only one species known from California *B. mirabilis*

BOLETUS MIRABILIS *Murrill, Mycologia 4:217. 1912*

Ceriomyces mirabilis Murrill, Mycologia 4:98. 1912.
Xerocomus mirabilis (Murrill) Singer, Rev. Mycol. 5:6. 1940.
Boletellus mirabilis (Murrill) Singer, Farlowia 2:129. 1945.

Illustrations: See Microfiche No. 13

Smith, A. H., Mushrooms in Their Natural Habitats, Reel V.
Smith, A. H., The Mushroom Hunter's Field Guide, p. 101.
Smith, A. H., and H. D. Thiers, The Boletes of Michigan, pls. 93, 94.
Snell, W. H., and E. A. Dick, The Boleti of Northeastern North America, pl. 3.

Pileus 7–15 cm broad, convex to obtusely convex when young, becoming broadly convex to plano-convex when mature; surface moist to lubricous when young, dry when older, somewhat appressed fibrillose-scaly to rough-tomentose when young, becoming conspicuously fibrillose-scaly and often squarrose-scaly with age, sometimes appearing granular-scaly to somewhat areolate or rimose-scaly; color dark reddish brown ("carob brown" to "bay" to "chocolate") unchanging during all stages of development, occasionally the scales slightly darker in color, pallid flesh rarely showing between the scales; margin sterile, often appendiculate, occasionally inrolled to incurved, becoming decurved with age. *Context* 1–2 cm thick, white to yellow ("naphthalene yellow" to "ivory yellow") often tinged vinaceous near the cuticle, frequently with numerous areas staining flesh-colored ("testaceous") when older, rarely changing to blue in irregular areas. Taste acid or mild; odor not distinctive.

Tubes 1–2.5 cm long, deeply and frequently broadly depressed around the stipe, yellow ("oil yellow" to "sulphur yellow") when young, changing to olive yellow ("olive lake") with age, unchanging when bruised; *pores* 1–2 mm broad, round to angular, concolorous with tubes, unchanging or becoming deeper yellow when bruised.

Stipe 8–15 cm long, 1–3 cm thick at the apex, clavate (stature of *B. russellii*), up to 5 cm thick at base, solid, with yellow mycelium at the base; surface moist to dry, glabrous, but typically shaggy reticulate at the apex, becoming smooth to uneven or ridged or shallowly pitted toward the base, sometimes becoming roughened and split with age; color dark brown ("carob brown" to "chocolate" to "bay") with occasional streaks of ivory or beige or brown. *Context* pale vinaceous ("avellaneous" to "testaceous"), sometimes appearing yellow and typically becoming yellow when dry.

Spore print olive brown. *Spores* 18–22 X 7–9 μm, ochraceous in KOH, dark brown in Melzer's, subfusoid to subellipsoid, smooth, thick-walled.

Basidia 31–36 X 7–11 μm, clavate, hyaline, four-spored. *Hymenial cystidia* 60–90 X 10–18 μm, scattered to numerous, hyaline, thin-walled, fusoid-ventricose with obtuse to elongated apices.

Tube trama gelatinous, hyaline, divergent from a central strand. *Pileus trama* homogeneous, interwoven. *Pileus cuticle* when young differentiated as a closely packed trichodermium, 120–160 μm thick, hyphae 10–15 μm wide, more or less cylindric, noticeably septate, terminal cells typically slightly tapered, pale yellow in KOH, reddish brown in Melzer's, with age unchanging or collapsing and becoming somewhat disarticulated. *Stipe cuticle* fertile in apical portion with scattered to somewhat clustered, large, thin-walled, hyaline, fusoid-ventricose caulocystidia, 50–80 X 12–18 μm, becoming loosely and irregularly interwoven toward the base. *Clamp connections* absent.

Chemical reactions NH_4OH—context very pale green; HNO_3—context pale vinaceous, cuticle pink.

Habit, habitat, and distribution Solitary to scattered in soil in coastal forests. Known only from the northern coastal counties. Occasionally found on rotten coastal hemlock logs or stumps. It never appears in great abundance in northern California, perhaps the southernmost extension of its range. In Oregon and Washington it is common, and typically found associated with hemlock logs and stumps.

Material studied Del Norte County: Thiers 14315, 17717, 21339, 21565. Humboldt County: Thiers 14503, 21382. Mendocino County: Thiers 10584, 10673, 14154, 14604, 21770, 24197, 27000; Wennekens 42.

Observations This is one of the most stately and easily recognized boletes in California. The uniform dark brown color, noticeably scaly pileus, and the long stipe are highly distinctive. In an earlier paper this species was considered to belong to the genus *Boletellus* because of its stature, general appearance, and because some workers had reported the spores as being obscurely punctate or roughened. Repeated examinations of California material have failed to reveal any roughened spores and since, in modern concepts *Boletellus* is restricted to species having such spores, it has been placed back in *Boletus*. Additional features showing affinity with *Boletus* are found in the divergent tube trama, yellow tubes, and dry to moist pileus surface.

This bolete is considered edible and choice.

Section Subtomentosi
SUBSECTION SUBTOMENTOSI
Key to Species

Couplet	Species	Go to
1. Ammonium hydroxide giving a fleeting but distinct blue to blue-green reaction on the surface of the pileus		2
1. Ammonium hydroxide not giving a blue or blue-green reaction, but may give some other color when applied to the surface of the pileus		3
2. Pileus yellow brown to olive brown; trama not amyloid	*B. spadiceus* var. *spadiceus*	
2. Pileus reddish brown; trama amyloid	*B. spadiceus* var. *rufobrunneus*	
3. Spores large (15–22 μm in length), often highly irregular in outline; pores pink to red	*B. mendocinensis*	
3. Basidiocarp not with the above characters		4
4. Pileus dull olive brown to yellow brown; pores 1–3 mm broad	*B. subtomentosus*	
4. Basidiocarp not with the above characters		5
5. Pileus dark gray brown to almost black		6
5. Pileus differently colored		7
6. Pileus fuscous to almost black; surface pruinose; cuticular hyphae irregularly incrusted; stipe often dark red with age	*B. zelleri*	
6. Pileus gray brown; conspicuously tomentose to velutinous, not pruinose; cuticular hyphae spirally incrusted; stipe pallid or red or reddish at apex	*B. chrysenteron*	
7. Stipe red or red and yellow, abruptly pinched at the base; pileus dry, tomen-		

tose to velutinous, brown to red or a mixture of the two colors; found under coastal live oak (*Q. agrifolia*) — *B. dryophilus*

7. Basidiocarp not as above; not found under coastal live oak — 8
    8. Pileus tan to buff, not red or reddish; tubes pale yellow; stipe yellow at the apex, red at the base, sometimes entirely red with age — *B. rubripes*
    8. Pileus red to reddish brown with olive or gray overtones; stipe red at the apex or with a red band in the apical region, often yellow or pallid at the base — *B. smithii*

BOLETUS CHRYSENTERON *Saint-Amans, Flore Agenaise, p. 555. 1821*

Xerocomus chrysenteron (Saint-Amans) Quelet, Fl. Mycol. Fr., p. 418. 1888.

Ceriomyces communis (Secretan, pro parte) Murrill, Mycologia 1:155. 1909.

Illustrations: See Microfiche No. 5

Kawamura, S., Icones of Japanese Fungi, pl. 235.
Lange, J. E., and M. Lange, 600 Pilze in Farben, p. 189.
Leclair, A., and H. Essette, Les Bolets, pl. 21.
Romagnesi, H., Nouvel Atlas des Champignons, pl. 123.
Singer, R., Die Rohrlinge, Teil I, pl. XII, figs. 5–12.
Smith, A. H., The Mushroom Hunter's Field Guide, p. 105.
Smith, A. H., and H. D. Thiers, The Boletes of Michigan, pl. 105.
Snell, W. H., and E. A. Dick, The Boleti of Northeastern North America, pl. 123.

Pileus 6–8 cm broad when expanded, convex changing to broadly convex to plano-convex to plane at maturity; surface dry, dull, velutinous to tomentose, usually shallowly and conspicuously areolate with age, especially near the margin; color dark brown ("Saccardo's umber" to "snuff brown") to occasionally blackish brown ("sepia" to "bister") during all stages of development; context in cracks on disc pallid, usually assuming reddish tints toward the margin when older; margin entire, decurved. *Context* 1–1.5 cm thick, yellow ("ivory yellow" to "light buff" to "pale chalcedony yellow"), unchanging or becoming blue in irregular areas when exposed. Taste not distinctive to somewhat acid and unpleasant; odor not distinctive.

Tubes 1 cm in length, arcuate-decurrent to depressed, often narrowly and deeply so in older carpophores, colored yellow ("reed yellow" to "old gold"), typically changing to blue when bruised; *pores* large (1–1.5 mm broad), typically highly irregular in outline, concolorous with the tubes.

Stipe 7–10 cm long, 1–1.7 cm thick at the apex, equal to tapering slightly toward the base, solid, whitish to yellow mycelium at the base; surface dry, glabrous but often longitudinally ridged or striate, usually yellow ("primrose yellow") at the apex becoming reddish ("russet" to "tawny" to "Hay's russet") toward the base, sometimes pallid at the base and reddish only in midportion, or entirely pallid with only the striations or ridges colored red

("Eugenia red"). *Context* pallid to yellowish at the apex, typically changing to reddish toward the base, unchanging when exposed.

Spore print olive brown. *Spores* 12–13.5 X 5–6 μm, ochraceous in KOH and Melzer's reagent, ellipsoid to subventricose to subcylindric, not truncate, smooth.

Basidia 33–36 X 7–9 μm, hyaline, four-spored, clavate. *Hymenial cystidia* 56–75 X 10–13 μm, scattered to numerous, common only on the pores, fusoid to fusoid-ventricose, hyaline, thin-walled.

Tube trama hyaline, obscurely divergent to subparallel, hyphae ±8 μm wide. *Pileus trama* homogeneous, interwoven, hyaline, hyphae ±7 μm wide. *Pileus cuticle* staining dark cinnamon brown in KOH, a trichodermium with slightly differentiated hyphal tips, hyphae ±10 μm wide, often heavily and spirally incrusted. *Stipe cuticle* interwoven, incrusted, staining dark cinnamon brown in KOH. *Clamp connections* absent.

Chemical reactions KOH—context negative to pinkish yellow, cuticle negative or darkening slightly; HCl—context negative to yellowish, cuticle negative to pinkish; HNO_3—context and cuticle negative to pale pink; sulfoformalin—context and cuticle negative to pale pink.

Habit, habitat and distribution Solitary to gregarious in soil in mixed forests. This bolete has been found in many different areas within the state, and probably occurs throughout the forested areas, including those of southern California. Apparently it forms mycorrhizal associations with a broad spectrum of hosts, since it has been found in both coniferous and broad-leaved forests. It is likely to be found more or less continuously, but never abundantly, from early in the fall season through February.

Material studied Butte County: Kowalski 1247. Del Norte County: Thiers 14108, 17639, 17745, 23037. El Dorado County: Thiers 20762, 20809. Humboldt County: Thiers 13748, 14544, 17615, 21385, 22909. Madera County: Thiers 20874, 20938, 23615. Marin County: Madden 694; Ripley 1003. Mendocino County: Largent 65; Theobald 11-2-63; Thiers 8746, 9294, 9470, 9617, 21524, 23065, 24166, 24440. San Diego County: Thiers 25195. San Mateo County: Thiers 9452. Santa Cruz County: Thiers 26970. Sierra County: Thiers 13850, 21182, 23636. Sonoma County: Sundberg 212. Stanislaus County: Thiers 12950, 16983. Tuolumne County: Thiers 21064.

Observations *Boletus chrysenteron* is a member of a rather difficult complex of boletes. It is distinguished by the dark-gray to grayish-black color of the dry, tomentose to velutinous pileus, the areolate condition of the mature pilei in which reddish pigments develop in the cracks, the yellowish to pallid stipe, which characteristically develops reddish areas with age, and the nontruncate spores. It is most likely to be confused with *B. truncatus,* from which it can be distinguished most readily by the absence of the characteristic truncate spores, and with *B. zelleri,* which usually has a noticeably pruinose but not strongly areolate pileus that is very dark colored, almost black, and a cuticle that, at least when young, is composed of a layer of somewhat inflated hyphal tips, which are not usually spirally incrusted. All three of these species often occur in the same area and are frequently quite difficult to separate.

Among other members of this complex are such common species as *B. subtomentosus,* which is usually more olive to olive brown in color with no red in the cracks, and *B. spadiceus,* which characteristically shows the blue to blue-green discoloration when ammonium hydroxide is applied to the cap. *Boletus amyloideus* and *Tylopilus amylosporus* can all be distinguished by their amyloid spores.

Edible but not choice, and rarely collected for the table. The possibility of confusing its identity with other, possibly untested, species should always be kept in mind.

BOLETUS DRYOPHILUS *sp. nov.*

Illustrations: See Microfiche No. 6

Pileus 4–10 cm latus, siccus, tomentosus vel velutinus, saepe rimosus, rufus vel rufobrunneus demum furvior vel marcidus. Contextus flavus tactu caeruleus. Sapor et odor mitis. Tubuli 1 cm longi, olivaceo-flavi tactu caerulei; pori 0.5 1 mm lati, concolores tactu caerulei. Stipes 5–8 cm longus, 1–2 cm crassus, siccus, glaber, apice flavus, rufus basim versus. Sporae 12–16.5 X 5.5–8 μm, subellipsoideae vel subfusoideae, laeves. Cystidia 25–40 X 7–12 μm, clavata. Cuticula intertexta cellulis clavatis vel vesiculosis gregatim. Holotypus (no. 18557) a H. D. Thiers lectus prope Fremont Campground, Los Padres National Forest, Santa Barbara County, January 28, 1967; in Herbarium San Francisco State University conservatus.

Pileus 4–10 cm broad when mature, convex to pulvinate when young, unchanging or becoming broadly convex to plane to highly irregular and undulating with age; surface dry to slightly moist, conspicuously tomentose when young, unchanging or appearing subglabrous with age, sometimes becoming rimose to split and often developing into large cracks or conspicuous fissures; color when young near reddish brown to red ("garnet brown" to "madder brown" to "Vandyke red" to "acajou red"), sometimes overlain by an olive brown ("tawny-olive" to pale "buckthorn brown") tomentum, with age, in some the red color darkens to near dark red ("Hay's maroon" to "Vandyke red" to "diamine brown"), in others the red fades and the olive brown ("tawny-olive" to "clay color") predominates; frequently, the combination of a red background and pallid tomentum gives the pileus a pinkish-brown appearance; margin incurved, becoming decurved, entire becoming eroded. *Context* 1–2 cm thick, yellow ("pale chalcedony yellow" to "sulphur yellow") with a slight reddening under the cuticle, slowly and erratically becoming blue when exposed. Taste slightly acid near the cuticle, otherwise mild; odor not distinctive.

Tubes up to 1 cm in length, broadly and deeply to shallowly depressed, sometimes appearing decurrent when very old, near olive yellow ("old gold" to "olive lake"), bluing when exposed; *pores* 0.5–1 mm broad, angular, uneven, concolorous with tubes, bluing when bruised.

Stipe 5–8 cm long, 1–2 cm thick at the apex, equal or enlarging slightly toward the apex, often pinched and narrowed at the base, solid; surface dry to moist, glabrous; color yellow at the apex ("barium yellow" to "citron yellow"), changing toward the base to reddish ("maroon" to "Vandyke red" to "diamine brown"), the red colors being particularly evident in old basidiocarps, darkening when bruised. *Context* concolorous with the surface, becoming blue when exposed except at the base.

Spore print brown. *Spores* 12–16.5 X 5.5–8 μm, subellipsoid to subfusoid, not truncate, smooth, moderately thick walled; brown in KOH, dark cinnamon brown in Melzer's.

Basidia 20–26 X 7–9 μm, hyaline, four-spored, clavate. *Hymenial cystidia* 25–40 X 7–12 μm, scattered, inconspicuous, and sometimes apparently absent, embedded in the hymenium, hyaline, clavate.

Tube trama parallel, becoming slightly divergent with age. *Pileus trama* interwoven, homogeneous. *Pileus cuticle* differentiated as a tangled trichodermium of heavily incrusted hyphae that stain pale ochraceous in KOH, the incrustations frequently spirally arranged. *Stipe cuticle* differentiated as a layer of interwoven hyphae with numerous clusters of clavate to vesiculose cells that are hyaline or brown in KOH. *Clamp connections* absent.

Chemical reactions HNO_3–context and cuticle red; HCl–context pink.

Habit, habitat, and distribution Solitary to gregarious in humus in open oak woods. This species has been found from the San Francisco Bay area south to San Diego County. In all instances it was associated with coastal live oak. It appears to be somewhat erratic in its fruiting habits, apparently not fruiting at all or only rarely during some seasons, but noticeably abundant in others. No doubt it will eventually be found to occur throughout the range of the coastal live oak.

Material studied Los Angeles County: Thiers 32189. Monterey County: Thiers 18595. San Diego County: Thiers 25066, 25131, 25214, 25223. San Francisco County: Thiers 24529. San Mateo County: Setzer 1513; Thiers 11921, 21902, 26952, 27015. Santa Barbara County: Thiers 18557 type, 25268.

Observations This species obviously belongs in the *Subtomentosi* section and also in the subsection *Subtomentosi.* It does not, however, satisfactorily fit the description of any species known for that taxon. It is recognized by the peculiar color combination in the pileus. The reddish color sometimes almost completely dominates, but more often the cuticle is red to rose colored when young but becomes overlain and at least partially masked with olive-brown fibrils as it gets older. The stipe is also distinctive in that it is rather short, often pinched at the base, and shows very distinct red and yellow portions. The cuticle is somewhat suggestive of that of *B. chrysenteron,* but the incrustations do not stain cinnamon brown as they do in that species. The colors of the pileus and stipe differentiate it from *B. subtomentosus,* and the absence of any blue discoloration on the pileus when ammonium hydroxide is applied precludes calling it *B. spadiceus.*

So far as is known, its edibility has not been determined.

BOLETUS MENDOCINENSIS *sp. nov.*

Pileus 7–10 cm latus, siccus, areolatus, tomentosus, fuscus, non rufus in fissuris. Contextus 1–1.5 cm crassus, alutaceus tactu immutabilis vel tarde caerulescens. Sapor et odor mitis. Tubuli 1 cm longi, flavi, tactu caerulei; pori 1–2 mm lati, rufi tactu caerulei. Stipes 10–11 cm longus, 1–1.5 cm crassus, siccus, appresso-fibrillosus, brunneus. Sporae 15–21 X 4–5 μm, anguste cylindricae. Cystidia 30–40 X 8–12 μm, conferta vel dispersa, clavata, apice elongata attenuata. Cuticulae hyphis intertextis et incrustatis composita. Holotypus (no. 18392) a H. D. Thiers lectus proper Jackson State Forest, Mendocino County, January 8, 1967; in Herbarium San Franciso State University conservatus.

Pileus 7–10 cm broad, convex when young, becoming plano-convex to plane with a flared or somewhat uplifted margin; surface dry, noticeably areolate, especially toward the disc, areoles small, tomentose to velutinous during all stages of development; color dark brown ("Saccardo's umber" to "snuff brown") to occasionally blackish brown ("fuscous" to "sepia" to "bister"), no reddish color apparent in the flesh exposed in the cracks on the pileus; margin entire, becoming somewhat rimose with age. *Context* 1–1.5 cm thick, tan ("warm buff" to "pale olive-buff"), unchanging or slowly changing to blue upon exposure. Taste and odor mild.

Tubes 1 cm in length, somewhat depressed, becoming subdecurrent, yellow ("reed yellow" to "old gold"), bluing when bruised or exposed; *pores* 1–2 mm broad, angular, red ("jasper red" to "dragons-blood red"), bluing when bruised.

Stipe 10–11 cm long, 1–1.5 cm thick at the apex, equal, solid to stuffed; surface dry, finely appressed-fibrillose to obscurely fibrillose-scaly; color brown ("cinnamon-brown" to "tawny-olive" to "sayal brown"), often with a reddish cast at the apex and base, brown color due primarily to the fibrillose covering, ground color pale yellow, yellow mycelium at the base. *Context* yellow at the base and apex, reddish in midportion, changing to blue when exposed.

Spore print color unknown. *Spores* 15–21(27) X 4–5(6) μm, narrowly cylindric in face view, slightly inequilateral in profile, giant spores and pleomorphism rather common, ochraceous in Melzer's (not dextrinoid), pale ochraceous in KOH; walls smooth, thin.

Basidia 24–29 X 7–9 μm, clavate, hyaline, four spored. *Hymenial cystidia* 30–40 X 8–12 μm, crowded on pores, scattered elsewhere, hyaline, thin-walled, clavate, apex often elongated or tapering.

Tube trama hyaline in KOH, obscurely divergent, hyphae 4–6 μm wide. *Pileus trama* homogeneous, interwoven. *Pileus cuticle* dark brown ("chestnut brown") in KOH, differentiated as a trichodermium of tangled hyphae with the terminal cell slightly enlarged, heavily incrusted but not spirally so. *Stipe cuticle* similar to that of the pileus. *Clamp connections* absent.

Chemical reactions unknown.

Habit, habitat, and distribution Solitary in dense hardwood–conifer forest. Until recently known only from the coast near Mendocino.

Material studied Mendocino County: Thiers 18392 type. Trinity County: Thiers 30659.

Observations This species obviously is in the *B. chrysenteron* complex, but is readily distinct because of the absence of any red pigment in the cracks on the pileus, the presence of the red pores, and the large spores. The absence of any amyloid reaction or truncate apex in the spores readily distinguishes it from *B. truncatus* and related species.

Edibility undetermined.

BOLETUS RUBRIPES *Thiers, Mycologia 57:532. 1965*

Illustrations: See Microfiche No. 17

Pileus 9–16 cm broad when mature, convex to obtusely convex when young, unchanging or becoming broadly convex with age; surface dry, dull, tomentose to velutinous to matted or appressed-fibrillose during all stages of development; color when young pale buff ("deep olive-buff" to "olive-buff"), unchanging or becoming pale brown ("buffy brown" to "tawny olive" to occasionally near "sayal brown" to "cinnamon") with age, sometimes fading to buff ("warm buff" to "light buff") on the margin, typically staining brown when bruised or handled; margin incurved to decurved, entire. *Context* up to 3 cm thick, buff to yellow ("pale olive-buff" to pale "ivory yellow" to "pale pinkish buff"), changing to blue immediately upon exposure. Taste typically unpleasant, bitter; odor often strong, unpleasant.

Tubes up to 1.5 cm in length, shallowly depressed when young, becoming broadly and deeply depressed with age; yellow ("primrose yellow" to "reed yellow") when young, changing to dark yellow ("massicot yellow" to "naphthalene yellow" to "chamois") with age, bluing immediately when bruised; *pores* up to 1 mm broad, angular, concolorous.

Stipe 8–13 cm long, 2–4 cm thick at the apex, equal to slightly clavate, solid, with yellow mycelium at the base; surface dry, glabrous or frequently appearing longitudinally striate, especially with age, not reticulate; color yellow ("primrose yellow") at the apex, changing to pink ("Daphne pink" to "rocellin purple") at the base, entire stipe typically becoming very dark red with age. *Context* yellow, dark red to almost black at the base, changing to blue when exposed.

Spore print olive brown. *Spores* 12.5–17.6 X 4–5 μm, ochraceous in KOH, dark ochraceous in Melzer's, smooth, thin-walled, subcylindric to subfusoid to narrowly subellipsoid.

Basidia 26–30 X 7–13 μm, four-spored, clavate, hyaline in KOH. *Hymenial cystidia* 24–35 X 5–7 μm, scattered, embedded in the hymenium, hyaline, thin-walled, fusoid to occasionally clavate or fusoid-ventricose.

Tube trama divergent to subparallel, hyaline, staining bright ochraceous in KOH near the pores. *Pileus trama* interwoven, homogeneous. *Pileus cuticle* differentiated as a trichodermium of interwoven hyphae, brown to pale ochraceous in KOH, occasionally incrusted. *Stipe cuticle* bright ochraceous in KOH, differentiated as a layer of closely packed hyphal tips that are septate, relatively thick walled, and superimposed upon a narrow layer of hyaline, interwoven hyphae that gelatinize in KOH. *Clamp connections* absent.

Chemical reactions KOH–context yellow orange; HNO_3–context yellow orange, cuticle pink; HCl–context yellow orange; sulfoformalin–context yellow, cuticle pink; guaiac–context pale blue.

Habit, habitat, and distribution Solitary to gregarious in humus in coniferous or mixed forests. It has not been possible to determine its mycorrhizal hosts in areas at lower elevations, but hardwoods such as oaks are suspected. In the Sierra and Cascades it has been found under both pines and firs. *Boletus rubripes* occurs most frequently in the coastal forests in the vicinity of Mendocino. Collections have been made, however, in most of the central and northern forests of the state. Although originally described from California, it is now known from throughout the Pacific Northwest and as far away as central Mexico.

Material studied Del Norte County: Sundberg 960. Mendocino County: Largent 566; Thiers 8926, 9266, 9299, 9352, 9435, 9473, 10580, 10589, 10619, 10675, 11593, 11618, 21317. Shasta County: Thiers 12726. Sierra County: Sundberg 1460; Thiers 13203. Tuolumne County: Sundberg 286; Thiers 13277.

Observations The pale-buff to tan color of the pileus and the smooth stipe, which, when young, is yellow at the apex and red at the base, but characteristically completely changes to dark red with age, are the major distinguishing features of this species. It appears, perhaps superficially, closely related to *B. calopus,* but the smooth stipe and pale color of the pileus make it readily distinct. Also, the dark red stipe of the older basidiocarps is not usually seen in *B. calopus.* The red stipe, the absence of any reticulation on its surface, and the bitter taste make it easily distinguishable from *B. appendiculatus,* a species not closely related but showing some similarity in surface characters.

Edibility unknown, but if the bitter taste persists after cooking, it would not be very palatable.

BOLETUS SMITHII *Thiers, Mycologia 57:530. 1965*

Illustrations: See Microfiche No. 19

Pileus 5–16 cm broad, convex when young, unchanging or becoming broadly convex to plano-convex when mature; surface dry, dull to unpolished, even when young but often with irregular depressions when older, velutinous to tomentose when young, becoming fibrillose to matted-fibrillose with age, sometimes appearing glabrous when very old and often becoming rimose-areolate upon drying or with age; background color olive yellow to buff ("pale olive-buff" to "deep olive-buff" to "tawny-olive" to "Isabella color"), but typically with pinkish or reddish overtones or blotches, when older frequently appearing red ("Pompeian red" to "jasper red") over entire surface or with the olive pigment confined to a narrow zone near the margin, sometimes becoming yellowish near the margin; margin entire, incurved, becoming decurved with age. *Context* 1–3 cm thick, pale yellow ("pale greenish yellow") except sometimes reddish near the cuticle, unchanging or slowly and erratically changing to blue when exposed, occasionally rapidly bluing. Taste and odor mild.

Tubes 1–1.5 cm long, adnate to arcuate decurrent when young, typically becoming deeply and narrowly depressed with age, yellow ("barium yellow" to "straw yellow") when young, becoming dark yellow ("reed yellow" to "olive-yellow") with age, bluing immediately upon exposure; *pores* 0.5–1 mm broad, angular, concolorous or sometimes appearing reddish near the margin when old, bluing when bruised.

Stipe 6–16 cm long, 1–3.5 cm thick at the apex, equal to tapering toward the apex or sometimes becoming clavate, solid; surface dry, smooth to granulose or pruinose, occasionally appearing appressed-fibrillose to tomentose, not reticulate; color red ("Eugenia red" to "Pompeian red" to "jasper red") toward the apex, changing to yellow ("naphthalene yellow" to "barium yellow") toward the base; sometimes the color pattern reversed or entirely red or, rarely, entirely yellow, usually with a distinct red band at the apex with age. *Context* yellow to pallid, bluing when exposed.

Spore print olive brown. *Spores* 14.5–19.2 X 4–6 μm, pale ochraceous in KOH, ochraceous to dark ochraceous in Melzer's, smooth, thin-walled, fusoid to subcylindric in face view, ventricose in side view.

Basidia 30–36 X 7–10 μm, four-spored, hyaline, clavate. *Hymenial cystidia* 30–60 X 7–12 μm, typically deeply embedded in the hymenium, sometimes obscure, hyaline, thin-walled, clavate to fusoid to fusoid-ventricose.

Tube trama hyaline to ochraceous in KOH, gelatinous, divergent from an obscure central strand, sometimes with occasional laticiferous hyphae. *Pileus trama* homogenous, interwoven. *Pileus cuticle* differentiated as a layer of loosely interwoven, tangled hyphae forming a poorly defined trichodermium, often with free hyphal tips, which appear somewhat cystidioid and sometimes fasciculate, some hyphae staining ochraceous in KOH, occasionally with numerous laticifers. *Stipe cuticle* differentiated as a layer of fertile basidia in the upper portion, changing to a layer of interwoven hyphae toward the base. *Clamp connections* absent.

Chemical reactions KOH—cuticle black; HCl—context pinkish yellow, cuticle pink; HNO_3—context pinkish yellow, cuticle reddish; sulfoformalin—context pinkish yellow, cuticle pinkish.

Habit, habitat, and distribution Solitary to gregarious in soil in dense mixed coastal forests. Curiously enough, this species has been found only in the vicinity of Mendocino in California, yet it was first collected in the Cascades in Oregon and has since been found in Washington and Idaho. It apparently fruits only during the fall—winter rainy season.

Material studied Mendocino County: Breckon 39; Motta 270; Thiers 8208, 9305, 9355, 9419, 9812, 10644, 10674, 11872, 14142, 14594, 21315, 21525, 23066, 24183, 24218.

Observations This species presents a very distinctive color pattern in the pileus in which there is a curious mixture of olive-gray, yellow, and red pigments. Most commonly, the background color is olive or olive brown or gray often overlain and sometimes obscured with red or pink overtones. With age, red frequently predominates, and the olive or gray pigments may completely disappear or become confined to a narrow marginal zone. The matted-fibrillose to tomentose or velutinous cuticle, the red band at the stipe apex, and the inconspicuous cystidia are additional distinctive features.

Edibility unknown.

BOLETUS SPADICEUS *var.* **SPADICEUS** *Fries, Epicr. Syst. Mycol., p. 415. 1838*

Xerocomus spadiceus (Fries) Quélet, Fl. Mycol., p. 417. 1888.

Suillus spadiceus (Fries) Kuntze, Rev. Gen. Pl. 3(2):536. 1898.

Xerocomus subtomentosus var. *spadiceus* (Fries) Gilbert, Les Bolets, p. 116. 1931.

Boletus subtomentosus ssp. *spadiceus* (Fries) Konrad and Maublanc, Icon. Sel. Fung. 6:463. 1938.

Xerocomus coniferarum Singer, Farlowia 2:297. 1945.

Illustrations: See Microfiche No. 20

Leclair, A., and H. Essette, Les Bolets, pl. 25.

Singer, R., Die Rohrlinge, Teil I, pl. XI, figs. 5–11.

Snell, W. H., and E. A. Dick, The Boleti of Northeastern North America, pl. 25.

Pileus 4–10 cm broad at maturity, convex to plano-convex during all stages of development; surface dry, dull, velutinous to subtomentose to tomentose during all stages of development; color when young brown to reddish brown ("Kaiser brown" to "russet" to as pale as "dark olive-buff"), with age usually becoming dark brown ("Kaiser brown" to "Dresden brown" to "cinnamon-brown") or fading to pale yellow brown ("Isabella color"), often retaining a reddish tint and appearing dark reddish brown when wet; margin incurved to decurved, entire. *Context* 1–1.5 cm thick, pale vinaceous near cuticle and yellow above the tubes, otherwise white, unchanging when exposed, or slowly changing to blue near the tubes. Taste and odor mild.

Tubes 1–1.5 cm in length, depressed to subdecurrent, yellow ("wax yellow" to "strontian yellow" to "mustard yellow"), changing to blue when bruised or exposed; *pores* 1–2 mm broad, often compound, angular, concolorous.

Stipe 5–15 cm long, 1–2 cm thick at the apex, equal to tapering toward the base, solid; surface dry, often coarsely reticulate in apical portion, smooth to longitudinally ridged toward the base; color pale yellow to buff ("ochraceous-buff" to "baryta yellow") to occasionally almost white, sometimes with brownish stains. *Context* white to pale yellow, unchanging or slowly changing to blue when exposed.

Spore print dark olive brown. *Spores* 9–11 X 4–5.5 μm, pale ochraceous in KOH and Melzer's, subfusoid to subellipsoid in face view, inequilateral and ventricose in profile, walls smooth, thin.

Basidia 20–30 X 4–6 μm, four-spored, hyaline, clavate. *Hymenial cystidia* 40–55 X 4–7 μm, numerous, conspicuous, projecting from the hymenium, clavate to fusoid to fusoid-ventricose, hyaline, thin-walled.

Tube trama obscurely divergent, hyaline, inamyloid, hyphae 6–7 μm. *Pileus trama* homogeneous, interwoven, inamyloid. *Pileus cuticle* differentiated as a trichodermium of suberect to tangled to repent hyphal tips, irregularly incrusted, ochraceous in KOH, rust red in Melzer's. *Clamp connections* absent.

Chemical reactions NH_4OH—cuticle blue, becoming fuscous; KOH—context orange yellow, cuticle dark brown.

Habit, habitat, and distribution Gregarious to solitary in soil in dense mixed coastal forests of northern California. It has also been collected in San Diego County.

Material studied Mendocino County: Thiers 24193, 26930. Napa County: Thiers 27005. San Diego County: Thiers 25234. San Mateo County: Thiers 26919, 26935, 26957, 26987, 26993, 27016. Santa Cruz County: Thiers 26968, 27052.

Observations This species is often confused with *B. subtomentosus*, but is most easily distinguished from it by the blue to blue-green discoloration of the surface of the pileus when ammonium hydroxide solution is applied. Also *B. spadiceus* often has more reddish pigment in the pileus and typically appears reddish brown when wet. Unfortunately, there is considerable variance in the concept of this species. Watling does not generally subscribe to the same species concept as that of most Americans and, as a matter of fact, specifically indicates that *B. spadiceus* does not show a blue discoloration with NH_4OH. He places fungi with this character in *B. lanatus*. However, until further data are available the present concept, which is accepted by both Smith and Singer, will be followed.

Edibility undetermined.

BOLETUS SPADICEUS *var.* RUFOBRUNNEUS *var. nov.*

Pileus 4–7.5 cm latus, siccus, subtomentosus, obscure rufobrunneus. Contextus albus tactu immutabilis. Tubuli 0.5–1 cm longi, flavi tactu caerulei; pori 1 mm lati, concolores tactu subcarnei. Stipes 4–6 cm longus, 1–1.5 cm crassus, siccus, glaber, subflavus. Sporae 11.5–15 X 5–5.5 μm, ellipsoideae vel subfusoideae. Cystidia 25–33 X 5–10 μm, fusoidea vel submucronata vel subclavata. Cuticula e hyphis rectis, clavatis, composita. Holotypus (no. 17766) a H. D. Thiers lectus prope Jedediah Smith Redwoods State Park, Del Norte County, November 12, 1966; in Herbarium San Francisco State University conservatus.

Pileus 4–7.5 cm broad at maturity, convex when young, becoming plano-convex to broadly convex at maturity; surface dry, dull, subtomentose to velutinous during all stages of development; color dark reddish brown ("auburn" to "Hessian brown" to "liver brown" to occasionally "chestnut-brown") during all stages of development; margin incurved, entire. *Context* 0.5–1 cm thick, white, unchanging when exposed. Taste and odor mild.

Tubes 0.5–1 cm in length, adnate to depressed, intensely yellow ("chamois" to "reed yellow"), unchanging with age, staining blue when exposed or bruised; *pores* up to 1 mm broad, appearing compound, concolorous with tubes, somewhat boletinoid, staining pale lavender to flesh color ("avellaneous" to "testaceous") when bruised.

Stipe 4–6 cm long, 1–1.5 cm thick at the apex, equal or tapering toward the base, solid, yellow rhizoids at the base; surface dry to moist, glabrous during all stages of development, pale yellow ("ivory yellow"), staining yellow brown or fulvous when bruised. *Context* white, unchanging when exposed.

Spore print color unknown. *Spores* 11.5–15 X 5–5.5 μm, pale ochraceous in KOH and Melzer's, ellipsoid to subfusoid in face view, inequilateral and ventricose in profile, not truncate, walls thin, smooth.

Basidia 21–27 X 5–8 μm, four-spored, clavate, giving a fleeting amyloid reaction when first mounted in Melzer's, soon becoming hyaline. *Hymenial cystidia* 25–33 X 5–10 μm, numerous, hyaline, pale ochraceous in Melzer's, fusoid to submucronate to subclavate, thin-walled.

Tube trama slightly divergent, giving a strong fleeting amyloid reaction when first mounted in Melzer's, quickly fading to hyaline, hyphae 4–5 μm wide. *Pileus trama* interwoven, homogeneous, nonamyloid, hyaline in KOH. *Pileus cuticle* differentiated as a trichodermium of erect, closely packed, clavate hyphal tips, heavily incrusted, not spirally arranged, incrustations staining bright red ("tawny") in Melzer's. *Clamp connections* absent.

Chemical reactions NH_4OH—cuticle greenish blue, changing to fuscous.

Habit, habitat, and distribution Gregarious in soil in dense mixed conifer–hardwood forest. It is known only from the coastal forest near Crescent City.

Material studied Del Norte County: Thiers 17766 type.

Observations This variety is distinguished by the reddish-brown color of the pileus and the amyloid reaction in the tube trama and hymenium. *Boletus spadiceus* var. *gracilis,* recently described from Michigan, has a similar amyloid reaction, but does not have a similarly colored pileus.

Edibility undetermined.

BOLETUS SUBTOMENTOSUS *Fries, Syst. Mycol., p. 389. 1821*

Xerocomus subtomentosus (Fries) Quélet, Fl. Mycol. Fr. p. 418. 1888.

Ceriomyces subtomentosus (Fries) Murrill, Mycologia 1:153. 1909.

Ceriomyces oregonensis Murrill, Mycologia 4:97. 1912.

Illustrations: See Microfiche No. 21

Kawamura, S., Icones of Japanese Fungi, pl. 237.

Lange, J. E., and M. Lange, 600 Pilze in Farben, p. 188.

Leclair, A., and H. Essette, Les Bolets, pl. 24.

Romagnesi, H., Nouvel Atlas des Champignons, pl. 124.

Singer, R., Die Rohrlinge, Teil I, pl. XII, figs. 13–15.

Smith, A. H., and H. D. Thiers, The Boletes of Michigan, pl. 104.

Snell, W. H., and E. A. Dick, The Boleti of Northeastern North America, pl. 26.

Pileus 6–15 cm broad at maturity, convex becoming broadly convex to plano-convex to plane or shallowly depressed at maturity; surface dry, dull, velutinous to subtomentose to matted tomentose during all stages of development, usually not conspicuously areolate but sometimes rimose at least on the margin; color when young yellow brown to olive brown ("Isabella color" to "tawny olive" to "clay color") to occasionally yellow ("yellow ocher" to "antimony yellow"), usually becoming darker brown ("sayal brown" to "snuff brown" to "buckthorn brown" to "olive-ocher") with age, no reddish pigmentation along the margin or in cracks in the flesh; margin entire, incurved to decurved, sometimes staining dark brown when handled. *Context* 1–2.5 cm thick, firm, whitish to pale yellow, often concolorous with the cuticle near the surface, typically changing to blue when exposed. Taste mild to slightly acid; odor not distinctive.

Tubes 1–2 cm long, adnexed to arcuate-decurrent when young, becoming shallowly to deeply depressed with age; when young bright yellow to buff ("antimony yellow" to "warm buff"), changing to deeper yellow ("primuline yellow" to "amber yellow" to "honey yellow" to "olive-ocher") with age, changing to blue when bruised; *pores* large, 1–2 mm, angular to highly irregular in outline, often compound, concolorous with tubes or rarely reddish.

Stipe 5–11 cm long, 1–2.5 cm thick at the apex, equal or sometimes narrowing at the base, solid, often with whitish to yellow mycelium and yellow rhizoids arising from the base; surface dry, glabrous to granulose to appressed fibrillose or longitudinally ridged, sometimes obscurely reticulate at the apex;

color yellow ("Naples yellow") at the apex, becoming rust colored ("tawny" to "ochraceous-tawny") in the midportion and pallid ("warm buff") at the base, occasionally with a yellow background color, but heavily overlain with darker-colored granulations. *Context* whitish to pale yellow, typically changing to blue when exposed.

Spore print olive brown. *Spores* 11.5–16 X 3.5–5 µm, pale yellow ochraceous in KOH, smooth, thin-walled, subfusoid to subcylindric.

Basidia 27–32 X 7–9 µm, clavate, hyaline in KOH, four-spored. *Hymenial cystidia* 40–66 X 8–12 µm, scattered to numerous, hyaline, thin-walled, fusoid to fusoid-ventricose, often with elongated, tapering apices.

Tube trama obscurely divergent to subparallel, with numerous laticiferous hyphae interspersed, hyphae ±5 µm wide. *Pileus trama* interwoven, homogeneous except for interspersed laticiferous hyphae. *Pileus cuticle* pale ochraceous yellow in Melzer's reagent and KOH, a more or less tangled trichodermium with numerous free, septate hyphal tips, usually strongly incrusted, hyphae 5–7 µm wide. *Stipe cuticle* differentiated as a layer of free to loosely interwoven hyphal tips with occasional tufts or clusters of caulocystidia 24–40 X 6–9 µm, which are hyaline, fusoid to clavate, thin-walled. *Clamp connections* absent.

Chemical reactions KOH—context negative to pale yellow, cuticle darkening slightly; NH_4OH—cuticle negative to darkening slightly; HCl—cuticle pink; HNO_3—context negative to pale yellow, cuticle dark red to pink; sulfoformalin—cuticle pink; $FeSO_4$—context negative or becoming pale gray.

Habit, habitat, and distribution Solitary to gregarious in soil in the coastal forests; often appearing along road cuts or similarly disturbed areas. It has only been found in abundance along the northern coast, but occasional collections have been made elsewhere within the state. Fruiting only during the fall–winter rainy season.

Material studied Del Norte County: Smith 55944; Sundberg 959; Thiers 14305, 17766. Humboldt County: Thiers 13974, 14190, 24148, 24305. Marin County: Madden 747; Thiers 7534, 10872, 11145. Mendocino County: Thiers 8180, 8396, 8405, 8876, 8877, 9061, 9262, 9292, 9309, 9337, 9348, 9438, 9471, 9767, 10573, 10647, 10676, 11019, 14595, 14615, 14616, 14619, 18403, 18496, 21772. San Mateo County: Sundberg 392; Thiers 8670, 10926, 11937, 18189, 21905. Santa Barbara County: Thiers 11798. Santa Cruz County: Thiers 8694, 10756, 10771, 10959, 10964, 18542. Siskiyou County: Thiers 23020.

Observations The dark-olive to yellow-olive color of the pileus, which is tomentose to velutinous, the bright-yellow tubes, which typically change to blue when bruised, the large, typically compound pores, and the more or less yellow stipe make this species distinct. In older, areolate pilei there is no appreciable development of red pigments in the cracks as in *B. chrysenteron.* The failure of ammonium hydroxide solution to give a blue discoloration on the cuticle of the pileus likewise distinguishes it from *B. spadiceus.* When compared with midwestern or eastern collections of *B. subtomentosus,* California material seems to be more heavily pigmented, particularly with olive tones, and to fade only slightly when dried.

An examination of the type of *Ceriomyces oregonensis,* described by Murrill from near Newport, Oregon, has revealed no significant differences from *B. subtomentosus,* and it is, consequently, reduced tc synonymy with it.

Edible and considered good by some.

BOLETUS ZELLERI *Murrill, Mycologia 4:217. 1912*

Ceriomyces zelleri Murrill, Mycologia 4:99. 1912.

Xerocomus zelleri (Murrill) Snell ex Slipp and Snell, Lloydia 7:43. 1944.

Boletellus zelleri (Murrill) Singer, Snell, and Dick, Mycologia 51:575. 1959.

Illustrations: See Microfiche No. 23

Smith, A. H., Mushrooms in Their Natural Habitats, Reel 6.

Smith, A. H., The Mushroom Hunter's Field Guide, p. 104.

Pileus 5–10 cm broad at maturity, convex to obtusely convex when young, becoming plano-convex to convex to plane with age; surface dry, conspicuously white-pruinose when young, becoming glabrous to obscurely tomentose to subtomentose with age, rugulose to verrucose when young, becoming more or less even or smooth at maturity, not conspicuously areolate when young, but sometimes becoming so with age; color black to blackish brown ("dusky brown" to "fuscous" to "fuscous-black" to "bone brown"), sometimes fading to dark brown ("army brown" to "Natal brown") often with a reddish margin; margin entire, becoming eroded with age, incurved when young, becoming decurved at maturity. *Context* up to 1.5 cm thick, whitish to pale yellow, unchanging or becoming blue when exposed. Taste not distinctive or mildly acid; odor not distinctive.

Tubes up to 1.5 cm in length, arcuate-decurrent to depressed, often deeply depressed with age, olive yellow ("olive-ocher") when young, changing to dark yellow ("honey yellow") with age, becoming blue when bruised or exposed; *pores* typically more than 1 mm broad, highly irregular in outline, concolorous with the tubes.

Stipe 5–8 cm long, 0.7–1.3 cm thick at the apex, equal to occasionally tapering slightly toward the apex, solid, with white to pale-yellow mycelium at the base; surface dry, granulose to punctate, especially toward the base, color tan ("warm buff") when young, becoming more or less obscured with red ("Eugenia red") granules, sometimes red ("acajou red") at the apex, becoming yellowish with red punctae toward the base, old stipes often red ("Vandyke red" to "madder brown"). *Context* yellow when young, typically red when older, sometimes changing to blue in irregular areas when exposed.

Spore print olive brown. *Spores* 12–15 X 4–5.5 μm, occasional giant spore measuring up to 24 μm in length, pale yellow to pale ochraceous in KOH, ochraceous tawny in Melzer's, subellipsoid to subventricose, smooth.

Basidia 18–21 X 10–11 µm, hyaline, clavate, four-spored, rarely one- to three-spored. *Hymenial cystidia* 40–85 X 10–13 µm, rare to scattered to numerous, apparently lacking in some carpophores, clavate to obtusely fusoid to fusoid-ventricose to obscurely mucronate, hyaline to rarely yellowish in KOH, thin-walled.

Tube trama parallel to obscurely divergent, with a distinct central strand, hyaline, hyphae ±6 µm wide. *Pileus trama* loosely interwoven, homogeneous, hyphae ±3 µm wide. *Pileus cuticle* staining dark brown in KOH, differentiated as a trichodermium of free, septate, erect, inflated hyphal tips, often appearing similar to pileocystidia, incrusted, typically collapsing and appearing as a tangled mass of hyphal tips in older pilei. *Stipe cuticle* interwoven, heavily incrusted, with laticiferous hyphae differentiated throughout. *Clamp connections* absent.

Chemical reactions NH_4OH–context greenish; HCl–context yellowish, cuticle dark pink; HNO_3–context and cuticle pink; sulfoformalin–context yellowish, cuticle dark pink.

Habit, habitat, and distribution Solitary to scattered in humus or soil in the coastal forests and in the Sierra and Cascade mountain ranges. Found most commonly in mixed coastal forests, especially where redwoods are relatively abundant. In some instances it has also been collected along the margin of slash burns and in restricted clearings within the forests. This is a common bolete in California and often fruits from the onset of the rainy period in the fall through March or April.

Material studied Butte County: Ripley 1638. Calaveras County: Sundberg 772. Del Norte County: Thiers 17640. Fresno County: Thiers 13367. Humboldt County: Sundberg 325; Thiers 14106, 14192, 17748, 19265. Marin County: Kady 11-23-68; Madden 677; Shervanick 382; Thiers 9825, 18148. Mendocino County: Ammirati 12-6-64; Motta 51; Peters 90; Thiers 8165, 8202, 8435, 8619, 8870, 8993, 9256, 9289, 9342, 9349, 9854, 10051, 10706, 11076, 12233, 14614, 18104, 24155, 24213. Nevada County: Thiers 13159. San Mateo County: Sundberg 196; Thiers 7449, 11199. Santa Cruz County: Peters 349; Thiers 9088. Shasta County: Thiers 21661. Sierra County: Sundberg 308; Thiers 13209. Sonoma County: Largent 324. Tuolumne County: Thiers 13275.

Observations This fungus appears to be very closely related to *B. chrysenteron*, and microscopic examination of the cuticle is often necessary to distinguish them. Generally, *B. zelleri* has a darker-colored pruinose pileus that typically does not become strongly rimose-areolate. It has been found, however, that by far the most reliable distinction is in the difference in the structure of the cuticle of the two species. In young pilei of *B. zelleri* the cuticle is differentiated as a closely packed trichodermium of noticeably enlarged erect hyphal tips in which the subterminal cell is globose to pyriform and the terminal cell is somewhat pyramidal in outline. In older pilei these cells sometimes collapse and the cuticle then appears repent to interwoven. In *B. chrysenteron*, on the other hand, the terminal hyphal cells are more or less equal in size and the hyphae are interwoven or more or less radially arranged. Another conspicuous feature of the cuticle of *B. chrysenteron* is the heavy incrustation on the hyphal walls, which is often deposited in a spiral pattern. The walls of the cuticular hyphae of *B. zelleri* are often incrusted, but not in a spiral fashion.

Singer, Snell, and Dick transferred *B. zelleri* to *Boletellus* chiefly because of their observation that the spores have slightly roughened walls. Careful attention, including the examination of the spores of local collections under oil with phase attachments as well as under oil with apochromatic objectives, failed to reveal any definite roughness or wrinkling of the spore wall. It is possible that their collections were from a different area, thereby representing a different race, or that the roughness has been overlooked in the local collections. It seems preferable to retain the species in *Boletus* and, if necessary, emend the concept to accommodate occasionally roughened spores.

Edible and generally considered excellent.

Section Subtomentosi
SUBSECTION VERSICOLORES
Key to Species

Only one species known from California *B. flaviporus*

BOLETUS FLAVIPORUS *Earle, Bull. N.Y. Botan. Garden 3:297. 1905*

Ceriomyces flaviporus (Earle) Murrill, Mycologia 1:147. 1909.
Pulveroboletus flaviporus (Earle) Singer, Am. Midland Nat. 37:17. 1947.

Illustrations: See Microfiche No. 11

Pileus 7–15 cm broad at maturity, convex when young, with age plano-convex to plane to somewhat depressed on the disc and flared to highly irregular on the margin; surface viscid to subviscid during all stages of development, tomentose or apparently appressed-fibrillose, sometimes appearing glabrous, often reticulate to irregularly rugose, rarely fibrillose-scaly on the margin with age; color when young reddish brown ("hazel" to "vinaceous-tawny" to "orange-cinammon"), with age darkening to dark reddish brown ("mahogany red" to "bay" to "chestnut") to sometimes dark brown ("seal brown" to "light seal brown") on the disc and fading to brown ("warm sepia" to "Verona brown" to "Mars brown") on the margin; margin entire, incurved, glabrous. *Context* 1–2 cm thick, white to pale pink ("pale pinkish buff"), unchanging when exposed or slowly becoming pinkish vinaceous ("testaceous" to "avellaneous"). Taste acid, somewhat unpleasant; odor mild.

Tubes 1–2 cm long, broadly and shallowly depressed, yellow ("yellowish citrine" to "pale green-yellow" to "light chalcedony yellow"), unchanging when exposed; *pores* 1 mm or more broad, angular, intense yellow ("strontian yellow" to "light greenish yellow" to "greenish yellow"), unchanging or darkening slightly when bruised.

Stipe 6–15 cm long, 1–3(4) cm thick at the apex, equal to clavate to tapering toward the base, solid, with conspicuous white mycelium at the base; surface viscid in wet weather, dry to subviscid otherwise, tubes forming a distinct, raised, yellow network at the apex, glabrous to punctate-roughened to occasionally fibrillose or fibrillose-scaly, sometimes ridged or obscurely reticulate; color yellow ("massicot yellow") at the apex, becoming white to pallid to nearly concolorous with the pileus, base typically white. *Context* white, unchanging.

Spore print dark olive brown. *Spores* 11–15 X 4–6 μm, pale ochraceous in KOH, cylindric to obscurely fusoid, smooth, thin-walled.

Basidia 23–29 X 8–12 μm, hyaline, clavate, four-spored. *Hymenial cystidia* 48–70 X 9–15 μm, scattered to numerous, hyaline, thin-walled, clavate to obscurely fusoid-ventricose, often appearing incrusted at the apex.

Tube trama hyaline, slightly divergent to subparallel. *Pileus trama* interwoven, homogeneous. *Pileus cuticle* differentiated as a pellicle of interwoven gelatinous hyphae, 3–4 μm wide. *Stipe cuticle* formed by a layer of interwoven, gelatinous hyphae with scattered areas in which the terminal cells are differentiated as enlarged clavate cells, 40–50 X 15–20 μm. *Clamp connections* absent.

Chemical reactions NH_4OH—context pale lavender; $FeSO_4$—context pale gray.

Habit, habitat, and distribution Solitary to scattered to occasionally cespitose in soil in mixed or hardwood forests. This bolete seems to be associated primarily with madrone, but occasional collections have been made under oaks. It has been found in the coastal area from Santa Barbara County northward to Mendocino County. The type was described from the Santa Cruz Peninsula, where it is abundant. Isolated collections have also been made in Napa, Sacramento, and Shasta counties. Surprisingly, no published reports have been found of its occurrence in other Pacific states. It fruits most commonly during the fall season, but single basidiocarps have been collected as late as April.

Material studied Marin County: Thiers 7523, 9827, 10871. Mendocino County: Thiers 21504. Napa County: Dusatko 7; Peters 325. Sacramento County: Thiers 32137. San Mateo County: Peters 156; Thiers 8546, 8547, 8663, 11818, 11918, 11965, 12173, 14652, 14837, 18063. Santa Barbara County: Thiers 18573. Santa Cruz County: Sundberg 201; Thiers 10772, 18549. Shasta County: Thiers 21606. Sonoma County: Thiers 9409, 10772.

Observations *Boletus flaviporus* is one of our most distinctive species and is easily recognized by the very intense yellow color of the pores, the viscid, cinnamon-colored pileus, and the pallid stipe, which is often punctate-roughened at least when young. Apparently, its nearest relative is *B. auriporus,* which is not known west of the Rocky Mountains. That species is much smaller, has a differently colored pileus, and a nonviscid stipe. In California, *B. flaviporus is* ***no***t likely to be confused with any other species.

The placement of this species in the genus *Pulveroboletus* seems to be indefensible. No evidence of a veil or pulverulence during any stage of development has been observed on any of the numerous collections examined. The fragments or appendiculations noted by Singer on old dried material could conceivably be artifacts resulting from the drying process.

Edibility unknown. Since *B. auriporus* is considered edible, this species is possibly safe, but, perhaps, not of a pleasing taste or texture.

SECTION TRUNCATI

Key to Species

1. Most spores conspicuously truncate; pileus brown to olive brown, tomentose, dry; spores not amyloid	*B. truncatus*
1. Only occasional spore with truncate apex; pileus red to reddish brown, glabrous, moist to subviscid (often viscid when wet); spores amyloid	*B. amyloideus*

BOLETUS AMYLOIDEUS *sp. nov.*

Illustrations: See Microfiche No. 3

Pileus 3–6 cm latus, uvidus vel subviscidus, glaber, rubidus vel obscure rufobrunneus. Contextus puniceus, immutabilis. Sapor et odor mitis. Tubuli 4–8 mm longi, subgalbani, tactu immutabiles; pori 1 mm lati, concolores tactu immutabiles. Stipes 8–10.5 cm longus, 1–1.5 cm crassus, siccus vel uvidus, glaber, subalutaceous vel alutaceus; contextus concolor tactu immutabilis. Sporae 13.5–15.5 X 4.5–5.5 μm, amyloideae, ellipsoideae vel fusoideae. Cystidia 30–48 X 7–15 μm, clavata vel subvesiculosa vel fusoideo-ventricosa. Cuticula e hyphis rectis composita. Holotypus (no. 26903) a H. D. Thiers lectus prope Jackson State Forest, Mendocino County, November 14, 1970; in Herbarium San Francisco State University conservatus.

Pileus 3–6 cm broad at maturity, convex when young, unchanging or becoming plano-convex to plane with age; surface moist to subviscid to viscid but never strongly so, glabrous during all stages, not rimose or areolate, evenly colored dark red to dark reddish brown ("Kaiser brown" to "dragons-blood red") when young, darkening to deep reddish brown ("Hay's russet") with age; margin entire, concolorous. *Context* 8–15 mm thick, pink ("alizarine pink" to "old rose"), unchanging when exposed. Taste and odor mild.

Tubes 4–8 mm in length, deeply to shallowly depressed, pale greenish yellow ("seafoam yellow" to "seafoam green"), not changing to blue when bruised; *pores* less than 1 mm broad, angular, bright greenish yellow ("lime green" to "reed yellow"), when bruised unchanging or darkening slightly or, less commonly, fading to pale greenish yellow ("seafoam yellow").

Stipe 8–10.5 cm long, 1–1.5 cm thick at the apex, clavate to subclavate, solid, surface dry to moist, not lubricous or viscid, glabrous during all stages of development; pale tan to tan ("warm buff" to "light buff"), sometimes with a reddish flush at the apex. *Context* white to concolorous with the surface, unchanging when exposed.

Spore print dark olive brown. *Spores* 13.5–15.5 X 4.5–5.5 μm, pale ochraceous in KOH, distinctly amyloid in Melzer's, ellipsoid to fusoid in face view, ventricose to inequilateral in profile, occasionally with truncate apex, walls smooth, thin.

Basidia 23–27 X 6–8 μm, hyaline in KOH, four-spored, clavate. *Hymenial cystidia* 30–48 X 7–15 μm, scattered, hyaline in KOH, clavate to subvesiculose to obscurely fusoid-ventricose, walls thin, not incrusted.

Tube trama hyaline in KOH, divergent, hyphae 5–6 μm wide. *Pileus trama* interwoven, homogenous. *Pileus cuticle* differentiated as a trichodermium of erect hyphal tips forming a turf, not gelatinizing in KOH, septate, terminal cells not at all or only slightly differentiated, not incrusted, pale yellow in KOH, bright red in Melzer's. *Stipe cuticle* similar to that of the pileus. *Clamp connections* absent.

Chemical reactions KOH—context white, cuticle dark brown; HCl—context of stipe yellow; $FeSO_4$—context gray.

Habit, habitat, and distribution Gregarious in soil in dense mixed hardwood–conifer forest. Known only from the coastal forests of Mendocino County.

Material studied Mendocino County: Cole 11-14-70; Thiers 26903 type.

Observations The discovery of this species adds another to the growing list of boletes with amyloid spores. Recently Smith described *Tylopilus amylosporus* from Northern Idaho, but it has a dark-reddish-brown spore deposit and at least some of the spores are truncate. Other species have been found to give at least fleetingly positive amyloid reactions in one or more parts of the basidiocarp. *Boletus amyloideus* is one of the few species known, however, in which all spores show this reaction and in which the reaction is strong enough to give a blue-black discoloration to the hymenium or to spore masses. It is further distinguished by the reddish color of the pileus, which is subviscid to viscid in wet weather, the bright-greenish-yellow tubes, and the failure of the context to change to blue when exposed.

Edibility unknown.

BOLETUS TRUNCATUS *(Singer, Snell, and Dick) Pouzar, Česká Mycol. 20:2. 1966*

Xerocomus truncatus Singer, Snell, and Dick in Snell, Singer, and Dick, Mycologia 51:573. 1959.

Illustrations: See Microfiche No. 22

Singer, R., Die Rohrlinge, Teil I, pl. XII, figs. 1–4.

Smith, A. H., and H. D. Thiers, The Boletes of Michigan, pl. 113.

Snell, W. H., and E. A. Dick, The Boleti of Northeastern North America, pl. 26.

Pileus 5–12 cm broad when mature, convex when young, becoming broadly convex to plano-convex to plane at maturity; surface dry, conspicuously tomentose during all stages of development, relatively smooth and even when young, usually becoming strongly areolate with age; color near brown ("tawny-olive" to "buffy brown") when young, darkening to dark brown ("clove brown" to "olive-brown" to "sepia" to "bister") when older, exposed flesh in cracks rarely showing reddish tints; margin entire, incurved. *Context* 1–1.5 cm thick, tan ("warm buff" to "light buff"), unchanging or becoming blue in irregular areas when exposed. Taste often acid and somewhat unpleasant; odor not distinctive.

Tubes 1–1.5 cm long, subdecurrent to deeply and narrowly depressed around the stipe, greenish yellow ("olive-ocher" to "Isabella color" to "old gold"), bluing readily when bruised; *pores* more or less 1 mm broad, angular, concolorous with tubes.

Stipe 8–12 cm long, 1.5–3 cm thick at the apex, equal to tapering slightly toward the base, solid, with bright-yellow mycelium at the base; surface dry, glabrous to obscurely punctate or ridged, color red ("hydrangea red" to "dark vinaceous") at the apex, becoming pallid ("warm buff" to "light buff" to "naphthalene yellow") toward the base, usually with red lines or ridges ("Eugenia red"). *Context* reddish toward the base, pallid at the apex.

Spore print olive brown. *Spores* 12.5–15 X 4.5–6.0 μm, tawny to cinnamon in KOH and Melzer's, ellipsoid to subventricose, conspicuously truncate, with thickened walls surrounding a thin depression in truncate apex, walls smooth.

Basidia 28–33 X 7–10 μm, hyaline, clavate, four-spored. *Hymenial cystidia* 36–67 X 10–14 μm, scattered to numerous, staining tawny to ochraceous in KOH, incrusted or filled with amorphous content, clavate to elongated fusoid-ventricose.

Tube trama hyaline, obscurely divergent to subparallel, hyphae ±6 μm wide. *Pileus trama* loosely interwoven, hyaline, homogeneous, hyphae 7 μm wide. *Pileus cuticle* differentiated as a trichodermium of interwoven to more or less radially arranged hyphal tips, often incrusted, staining ochraceous to tawny in KOH, hyphae ±10 μm wide. *Stipe cuticle* differentiated as a layer of caulobasidia with numerous sterile hyphal tips interspersed, staining cinnamon to tawny in KOH, tawny laticiferous hyphae scattered throughout. *Clamp connections* absent.

Chemical reactions HCl–context yellowish; sulfoformalin–context pink.

Habit, habitat, and distribution Solitary to scattered in humus in mixed forests. It is commonly found along the coast and has been collected from Santa Barbara to the Oregon border. Only one collection has been made elsewhere in California and that was in the vicinity of Paradise in Butte County. It is widely distributed throughout most of the forested regions of the United States. On the west coast it typically fruits during the fall and winter rainy season.

Material studied Butte County: Ripley 1638. Del Norte County: Thiers 17638. Humboldt County: Thiers 14439, 23014; Wennekens 154; White 211. Marin County: Madden 815; Thiers 11134, 18119. Mendocino County: Ammirati 12-6-64; Shamiyeh 11-2-63; Thiers 8203, 14617, 24443. San Francisco County: Thiers 24721. San Mateo County: Motta 12-6-60; Thiers 11917, 18061, 18195, 18355, 21847, 26931, 26991, 27017. Santa Barbara County: Thiers 11797. Santa Cruz County: Thiers 10786, 12038, 17935, 17961, 18532, 26969. Sonoma County: Thiers 9412.

Observations When first seen, this bolete is likely to be determined as *B. chrysenteron* since they are similar in color and, in both, the pileus typically becomes areolate with age. *Boletus truncatus,* however, does not usually show the characteristic red or reddish pigment in the cracks as does *B. chrysenteron.* It should be emphasized that a microscopic examination is necessary, however, before a positive determination of *B. truncatus* can be reliably accomplished. The noticeably truncate apex of at least a majority of the spores is its major distinguishing characteristic. Macroscopically, *B. truncatus* is sometimes confused with *B. zelleri,* which, however, has a distinct pruinose covering on the pileus and is usually darker in color. The cells of the terminal hyphae of the cuticle of *B. zelleri* are more inflated and the spores are not truncate.

Edibility not determined.

GASTROBOLETUS LOHWAG

Beih. Bot. Centralbl. 422:273. 1926

"Pileus" up to 8 cm broad, generally convex but typically highly irregular in outline and shape, especially with age, sometimes only partially coming above ground; surface dry to moist to subviscid, glabrous to tomentose to appressed fibrillose, sometimes appressed fibrillose-scaly, white to pallid to yellow brown to reddish brown to red; margin incurved, often highly irregular (peridium extending from margin over most of the gleba in *G. subalpinus*); context 1–2 cm thick, white to yellow, unchanging or becoming pinkish or blue when exposed; taste and odor typically mild. "Tubes" 1–2 cm long, highly irregular in arrangement, often not vertically oriented, white to yellow, unchanging or changing to blue when bruised; pores up to 1 mm broad, angular, concolorous to sometimes red or reddish. Stipe-columella typically present, often poorly developed and sometimes eccentric in attachment. Spore print not obtainable; spores ellipsoid to oblong to subovoid, ochraceous in KOH, smooth, walls thin to moderately thick. Hymenial cystidia scattered to abundant, sometimes clustered, clavate to fusoid to fusoid-ventricose, staining dark brown to black in KOH in at least one species. Tuba trama divergent to parallel. Cuticle a trichodermium, sometimes approximating a cutis. Clamp connections absent.

Gastroboletus Lohwag

Key to Species

1. Peridial membrane white, persistent, enclosing the gleba and well differentiated from the cuticle of the cap	*G. subalpinus*	
1. Gastrocarp not with the above characters		2
2. At least 50 percent of the spores truncate and with a small germ pore	*G. xerocomoides*	
2. Spores not as above		3
3. Hymenium with clustered, darkly staining cystidia when mounted in potassium hydroxide	*G. suilloides*	
3. Hymenium without clustered cystidia		4
4. Context not turning blue when exposed; spores noticeably amyloid	*G. amyloideus*	
4. Context bluing rapidly when exposed; spores not amyloid	*G. turbinatus*	

GASTROBOLETUS AMYLOIDEUS *Thiers, Brittonia 21:251. 1969*

Gastrocarp 4–8 cm broad at maturity, convex to broadly convex when young, becoming flattened to depressed to plano-convex at maturity; surface dry, dull, glabrous to subtomentose during all stages of development; color yellow ("colonial buff" to "old gold") to buff or rust brown ("ochraceous-buff" to "ochraceous-tawny"), or with reddish areas scattered throughout. *Context* 1–2 cm thick, pale yellow, unchanging when exposed. Taste and odor not distinctive.

Gleba tubulose, tubes 1–2.5 cm long, highly disoriented and appearing to radiate at various angles from the peridium, characteristically yellow ("colonial buff" to "chamois") during all stages of development, unchanging when bruised or exposed; *pores* small, less than 1 mm broad, concolorous with the tubes to occasionally reddish.

Stipe-columella 1–2 cm long, 0.5–1.5 cm thick, very reduced and sometimes apparently lacking, solid; surface dry, dull, glabrous, yellow ("chamois" to "old gold" to as pale as "warm buff") during all stages of development, with a characteristic red band at the apex. *Context* yellow, unchanging upon exposure.

Spores 13.5–18.8 X 6–7 μm, brown in KOH, strongly amyloid when seen in mass in Melzer's, fusoid to subcylindric, thick-walled, smooth, sterigmata terminal.

Basidia 26–32 X 7–10 μm, hyaline, clavate, four-spored. *Hymenial cystidia* apparently not present.

Tube trama hyaline, weakly divergent, at least in young gastrocarps. *Peridial trama* up to 15 μm broad, interwoven, septations in Melzer's becoming conspicuous and giving either a dextrinoid or amyloid reaction. *Cuticle of upper peridium* interwoven to tangled, pale ochraceous to dark yellow in KOH and Melzer's, hyphae 5–7 μm wide. *Clamp connections* absent.

Habit, habitat, and distribution Hypogeous under lodgepole pine and red fir in higher elevations in Sierra Nevada in California. Collected in late summer.

Material studied Amador County: Thiers 32641. Sierra County: Thiers 21117, 32926, 32927.

Observations As the name indicates, the most striking characteristic of this fungus is the conspicuous amyloid reaction of the spores and the differential staining of the hyphal septations. Both amyloid reactions are typically strong enough to be seen clearly with the low-power objective of the microscope. It is interesting to note that a similar staining of the septa has been observed in *Boletus calopus.* In most other features this fungus is somewhat suggestive of *G. turbinatus,* although the flesh did not change to blue.

Edibility undetermined.

GASTROBOLETUS SUBALPINUS *Trappe and Thiers, Brittonia 21:245. 1969*

Illustrations: See Microfiche No. 25

Gastrocarp 6–10 cm broad, convex, becoming plane to deeply depressed to plano-convex at maturity, frequently highly pitted and wrinkled and irregular in outline; surface dry to moist, glabrous to velutinous to subtomentose, typically covered with adherent soil and debris; color when young pale tan ("warm buff" to "light buff" to "pale olive-buff" to "pinkish cinnamon"), unchanging with age or darkening to near pale brown ("clay color" to "buckthorn brown"), peridium on lateral and undersides of gleba white, thin, velvety, persistent except for scattered, circular ruptures over mouths of occasional glebal tubes in age. *Context* 0.3–1.5 cm thick at disc, soft, white or sometimes with sordid yellowish to olivaceous zones, in some slowly staining pinkish to grayish lavender when cut. Odor mild to farinaceous; taste mild, pleasant.

Gleba tubulose, the tubes 1–3 cm long, but markedly diminished in length in both directions from the broadest point of the hymenophore; grayish yellow ("dark olive-buff" to "pale olive-buff"), with age becoming olive brown ("buffy citrine" to "citrine-drab"). Gleba depressed to free from stipe-columella, strongly projecting beyond upper peridium but then covered by the thin, pallid, lateral peridium. Tubes occasionally oriented vertically near the stipe-columella, but mostly curved and oriented 20 degrees from vertical to horizontal, or angled upward near edges of the hymenophore; *pores* small, less than 1 mm broad, concolorous with tubes, unchanging when bruised.

Stipe-columella 2–5 cm long, 2–4.5 cm thick at apex, subventricose or tapering downward, occasionally equal to bulbous at the base, base up to 3–4 cm broad in some gastrocarps; surface dry, but typically covered with adherent soil and debris, glabrous to velutinous to subtomentose with the apex reticulate when the gleba has seceded; color dirty white to tan ("warm buff" to "light buff"), in some the base orange yellow ("apricot yellow"). *Context* white, unchanging.

Spores 10–16(18) × 4.5–6(8) μm, smooth, ellipsoid, oblong to subovate, occasionally irregularly constricted near middle, light yellow to ochraceous in KOH and Melzer's, moderately thick walled, sterigmal appendages oblique.

Basidia 40–48 × 8–13 μm, thin-walled, hyaline and guttulate in KOH, clavate, with four sterigmata, 3–4 μm long. *Hymenial cystidia* 24–75 × 7–10 μm, often abundant when young, especially on the pores, apparently collapsing with age, single or sometimes clustered, hyaline in KOH and Melzer's, thin-walled, subcylindric to clavate to tapered.

Tube trama 60–80 µm wide, consisting of hyaline, parallel hyphae 4–6(8) µm wide. *Peridial trama* of thin-walled, often inflated, loosely interwoven hyphae (3)8–25 µm wide, hyaline laticiferous hyphae 12–20 µm wide with walls thickened to 1 µm scattered throughout. *Peridial cuticle* on upper surface of gastrocarp composed of closely interwoven, much branched hyphae 4–15 µm wide, with erect, tapered, blunt-tipped terminal cells 25–45 X 7–12 µm, sparsely scattered over the surface, hyaline in KOH, in Melzer's pallid to pale yellow to bright reddish brown, with darkly colored areas having intercellular deposits of dextrinoid debris. Peridial cuticle over lateral and undersides of gleba when young a trichodermium of blunt-tipped, cylindric to tapered to occasionally subcapitate cells 27–65 X 8–12 µm, collapsing with age. *Cuticle of stripe-columella* when young a palisade of hyaline clavate cells 22–35 X 6–15 µm, with occasional fertile basidia, collapsing with age. *Clamp connections* absent.

Habit, habitat, and distribution Solitary to gregarious, hypogeous or partially erumpent, under and probably mycorrhizal with pines and firs and possibly other conifers of montane and subalpine forests. Commonly found in late spring and summer.

Material studied Calaveras County: Thiers 21266, 21270. Fresno County: Thiers 13422, 13436. Sierra County: Thiers 21151, 23955, 23956, 23647. Tuolumne County: Thiers 21017, 21071, 21292.

Observations *Gastroboletus subalpinus* is readily distinguished from the other known species of the genus by its smooth, generally whitish pallid peridium and stipe. It is distinctly gastroid in having a gleba whose tubes are not vertically oriented, and mostly covered by peridial tissue through maturity.

When first collected in California, the specimens were thought to be abnormal basidiocarps of *Boletus edulis,* although the tubes were more or less enclosed and disorganized in orientation. Subsequent collections made it apparent, however, that these were basidiocarps of a species of *Gastroboletus.* In the case of each collection, careful search was made for normal-appearing basidiocarps of *B. edulis,* but none was found.

Edibility undetermined.

GASTROBOLETUS SUILLOIDES *Thiers, Brittonia 21:248. 1969*

Gastrocarp 2.5–5 cm broad at maturity, convex to subglobose to plano-convex during all stages of development; surface dry, glabrous when young, often becoming fibrillose to tomentose to appressed fibrillose-scaly with age, sometimes becoming areolate with age; color light brown ("ochraceous-buff" to "cinnamon-brown" to "buckthorn brown") to occasionally brown to dark brown ("mikado brown" to "sayal brown") when older, unchanging when bruised. *Context* 3–7 mm thick, buff to yellow ("ochraceous-buff" to pale "cream color"), unchanging when exposed. Taste and odor not distinctive.

Gleba distinctly tubulose, tubes arranged so as to make a spore deposit impossible, typically exposed at maturity, greenish yellow to olive yellow, unchanging when bruised or exposed; *pores* small, irregular, concolorous with the tubes.

Stipe-columella typically poorly developed, 5–10 mm long, 3–5 mm thick, surface dry, tomentose to appressed-fibrillose, more or less concolorous with the pileus during all stages of development.

Spores 6.4–12 X 3.5–4.5 μm, brown to dark brown in KOH and Melzer's, ellipsoid to inequilateral, some appearing ovoid or occasionally more or less constricted at the midportion, smooth, walls slightly thickened, sterigmal appendages oblique.

Basidia 20–37 X 6–8 μm, narrowly clavate to subcylindric, clavate to depressed at the center, hyaline in KOH, with four sterigmata. *Hymenial cystidia* 35–55 X 5–11 μm, abundant to numerous, typically clustered, and staining dark brown in KOH and Melzer's (as in *Sullius*), clavate to cylindric to subcylindric.

Tube trama divergent from a distinct central strand, hyphae 4–9 μm broad. *Peridial trama* interwoven, hyphae 5–30 μm broad, staining bright ochraceous in KOH, deep orange in Melzer's. *Peridial cuticle* appressed, of interwoven to subparallel to sometimes tangled hyphae, 5–12 μm broad, yellow to dark brown in KOH and Melzer's, not gelatinous. *Cuticle of stipe-columella* composed of loosely subparallel hyphae 7–30 μm broad and having some inflated cells up to 50 μm broad, grayish yellow in KOH. *Clamp connections* absent.

Habit, habitat, and distribution Solitary to gregarious, hypogeous to partially emergent, under conifers, including red fir and lodgepole pine. Known only from higher elevations in the Sierra Nevada of California in late summer.

Material studied Fresno County: Thiers 13412. Inyo County: Thiers 20990. Tehama County: Cooke 38731.

Observations This species shows close affinity with the genus *Suillus* because of the presence of fascicled cystidia that darken in KOH. This relationship is further substantiated by the similarity of shape and size of the spores of the two taxa. However, neither a viscid pellicle nor an annulus was observed in any collections of *G. suilloides.*

Edibility undetermined.

GASTROBOLETUS TURBINATUS *(Snell) Smith and Singer, Brittonia 11:208. 1959*

Boletus turbinatus Snell, Mycologia 28:20. 1936.

Boletellus turbinatus (Snell) Singer, Farlowia 2:134. 1945 withour citation of basionym. Lilloa 22:696. 1951.

Illustrations: See Microfiche No. 26

Gastrocarps 3–5 cm broad, convex, becoming highly irregular in outline, typically boletoid in appearance; peridium dry, relatively thin, surface with numerous pits and wrinkles, tomentose to velutinous; color brown ("mummy brown" to "chestnut-brown"), characteristically with reddish or yellowish tints or spots with age; margin incurved to decurved, highly irregular. *Context* 3–7 mm thick, yellow, changing to blue when exposed. Taste and odor mild.

Gleba tubulose, tubes 1.5–3 cm in length, highly irregular and sinuous in arrangement, never truly vertical, broadly depressed around the stipe columella; color yellow ("greenish yellow"), changing to blue when bruised; *pores* 0.5–1 mm broad, angular, typically pink to red, bluing when bruised.

Stipe-columella 2–4 cm long, 1–2 cm thick at the apex, typically equal, except narrowed at the base, often not extending much beyond the gleba, solid; surface dry, yellow but often covered with small reddish scales or granules, glabrescent. *Context* yellow, changing to blue when bruised.

Spores 13.5–15.5 X 7.5–8.5 μm, fusoid to subellipsoid in face view, slightly inequilateral in profile, ochraceous in KOH, walls 0.5–1 μm thick, smooth, often with slight germ pores.

Basidia 25–37 X 8–12 μm, hyaline in KOH, guttulate, clavate, four-spored. *Hymenial cystidia* 40–58 X 8–15 μm, uncommon, inconspicuous, clavate to cylindric to fusoid, hyaline, thin-walled.

Tube trama obscurely divergent, hyaline, hyphae 3–5 μm wide. *Peridial trama* hyaline, interwoven, homogeneous. *Peridial cuticle* differentiated as an appressed to tangled layer of hyphae, 4–5 μm wide, contents staining bright brown in KOH, walls not incrusted. *Clamp connections* absent.

Habit, habitat, and distribution Scattered to gregarious in soil under conifers. In this state it is known only from higher altitudes, where it appears to be associated with red and white fir.

Material studied Calaveras County: Thiers 21252, 21256. Lassen County: Kowalski 3835. Sierra County: Thiers 23648, 23649, 23650, 23652, 23960. Siskiyou County: Cooke 38660, 38661.

Observations This was the first gastrobolete known from North America and appears to be widely distributed throughout the temperate conifer forests of the continent. It is recognized by the brown cap with yellow to red spots or flushes, the consistent blue discoloration of the context when bruised, and the large, thick-walled spores, which are neither truncate nor amyloid.

Edibility undetermined.

GASTROBOLETUS XEROCOMOIDES *Trappe and Thiers, Brittonia 21:247. 1969*

Gastrocarp 30–60 mm broad, convex to plano-convex; peridium dry, tomentose, occasionally somewhat rimose, suggestive of *Boletus subtomentosus;* color olive yellow ("Isabella color" or "tawny-olive" to as dark as "Saccardo's umber"); margin incurved, entire. *Context* up to 1.5 cm thick at disc, whitish but staining brown in some areas and blue in others when exposed.

Gleba tubulose, the tubes up to 1.5 cm long, not vertically oriented, deep yellow ("old gold"), staining blue where cut or bruised; *pores* concolorous but staining brownish to ochraceous where bruised.

Stipe-columella poorly developed, 1–2.5 cm long, 0.4–0.8 cm broad, occasionally much shorter or lacking, equal or tapered downward; surface dry, glabrous to furfuraceous, pallid to pale tan ("warm buff"), staining brown with age or where bruised. *Context* pallid at apex, grading to distinctly reddish at the base, staining blue in irregular areas when exposed.

Spores 12.8–18(21) X 6.4–8 μm, smooth, ellipsoid to subcylindric, inequilateral, pyriform, or slightly constricted in the middle, about 50 percent being truncate, yellowish in KOH.

Basidia 30–38 X 10–12 μm, clavate, yellow in KOH, with one to four sterigmata, hyaline. No hymenial cystidia noted.

Tube trama of parallel to subparallel hyphae 4–8 μm broad; hyaline to pale yellowish in KOH and light yellow in Melzer's. *Peridial trama* of interwoven hyphae 3–10 μm wide; hyaline to pale yellowish in KOH, in Melzer's pale to deep reddish brown to dull yellow. *Peridial cuticle* a trichodermium of erect hyphal tips with subcylindrical to tapered terminal cells, hyphae 5–10 μm wide; golden brown in KOH, orange brown to reddish brown in Melzer's; hyphae irregularly incrusted with bands or patches of golden-yellow material that slowly dissolves in KOH or Melzer's. *Cuticle of stipe-columella* differentiated as a hymenium of fertile basidia. *Clamp connections* absent.

Habit, habitat, and distribution Gregarious in humus in coniferous forests usually in the vincinity of red fir at higher elevations in the Sierra Nevada in late summer.

Material studied Sierra County: Yuba Pass, Thiers 13163, 21116.

Observations This gastrobolete, showing obvious affinities with the xerocomoid group of boletes, is the only known species with truncate spores. The structure of the apical pore is similar to that in *Boletus truncatus.* The erratic bluing of the context and the dry tomentose peridium are macroscopic characters similar to species formerly placed in *Xerocomus.* Microscopically, similarities are seen in the structure of the cutis, including the incrustations on the cuticular hyphae.

Edibility undetermined.

GYROPORUS QUÉLET

Enchir. Fung. p. 161. 1886

Pileus up to 12 cm broad, convex to broadly convex becoming plane to shallowly depressed to plano-convex with age; surface dry to moist, glabrous to tomentose to appressed fibrillose to occasionally squamulose or woolly, pallid to buff to tan to brown; margin entire, incurved; context white, unchanging or changing to blue, sometimes rapidly and intensely so (context unchanging in *G. castaneus*); taste and odor mild. Tubes up to 8 mm long, white to very pale yellow when young, becoming bright yellow with age, unchanging or becoming blue in some when bruised; pores round to obscurely angular, one to two per millimeter, concolorous. Stipe up to 10 cm long, centrally attached, equal to clavate, typically hollow, at least in basal portion with age; surface glabrous to tomentose to appressed fibrillose (no annulus or veil in *G. castaneus*), pallid to buff to cinnamon to brown, unchanging or changing to blue when bruised. Spores yellow in deposit, hyaline in KOH, short ellipsoid, smooth, thin-walled. Hymenial cystidia abundant, hyaline, thin-walled, fusoid to fusoid-ventricose. Tube trama divergent. Cuticle a trichodermium or cutis. Hyphae of stipe transversely or horizontally arranged. Clamp and false clamp connections present.

Gyroporus Quélet

Key to Species

Only one species known from California *G. castaneus*

GYROPORUS CASTANEUS *(Fries) Quélet, Enchir. Fung., p. 161. 1886*

Boletus castaneus Fries, Syst. Mycol. 1:392. 1821

Illustrations: See Microfiche No. 24

Kawamura, S., Icones of Japanese Fungi, pl. 248.
Leclair, A., and H. Essette, Les Bolets, pl. 70.
Romagnesi, H., Nouvel Atlas des Champignons, pl. 137.
Singer, R., Die Rohrlinge, Teil I, pl. II, figs. 1–4.
Smith, A. H., The Mushroom Hunter's Field Guide, p. 75.
Smith, A. H., and H. D. Thiers, The Boletes of Michigan, pls. 2–4.
Snell, W. H., and E. A. Dick, The Boleti of Northeastern North America, pl. 7.

Pileus 3–7 cm broad, obtuse to convex when young, becoming broadly convex to plano-convex to plane or depressed with age; surface dry, glabrous to finely tomentose to pulverulent, often rimose or areolate-rimose with age, colored brown to tan to pinkish brown ("Mikado brown" to "orange-cinnamon" to "pinkish cinnamon" to "pecan brown" to "antique brown"); margin entire, but becoming rimose or split, incurved, becoming decurved or flared. *Context* 5–7 mm thick, white, unchanging when exposed; taste and odor mild.

Tubes 3–5 mm long, shallow to deeply and broadly depressed around the stipe, white when young, becoming yellow ("Naples yellow" to "barium yellow" to "amber yellow") at maturity, unchanging or darkening slightly when bruised; *pores* one to two per millimeter, angular, radially elongate near stipe apex, concolorous with tubes.

Stipe 2–4 cm long, 0.5–1.5 cm thick at the apex, equal to tapering toward the apex, hollow in basal portion; surface dry, unpolished, glabrous to tomentose to velutinous, colored similarly to the pileus—tan to brown to pinkish brown. *Context* white, unchanging.

Spore print yellow. *Spores* 10–12.5(14) X 5–6 μm, hyaline to pale yellow in KOH, ochraceous in Melzer's, ellipsoid, inequilateral in side view, smooth, thin-walled.

Basidia 30–35 X 8–14 μm, four-spored, clavate, hyaline to pale yellow in KOH. *Hymenial cystida* 30–45 X 7–11 μm, abundant, hyaline, thin-walled, fusoid to obscurely fusoid-ventricose.

Tube trama hyaline, divergent. *Pileus trama* nonamyloid, interwoven, hyphae 6–8 μm. *Pileus cuticle* differentiated as a layer of cystidioid cells forming a tangled trichodermium. *Stipe cuticle* more or less similar to the pileus cuticle, hyphae of context transversely oriented. *Clamp connections* present.

Chemical reactions Singer reports the following data: Metol–positive; phenol–chocolate brown; lactic acid–cuticle purple red; guiaic–context slowly and weakly blue green.

Habit, habitat, and distribution Gregarious in sandy alluvium along the delta of the Sacramento River. So far it has been found only under oaks. Although relatively common in the eastern half of the United States, this is apparently a very rare bolete on the west coast. It is known from only a few collections in California. These were made in the same place and the basidiocarps probably developed from the same mycelium. A few collections have also been made in Oregon and Washington.

Material studied Solano County: Sundberg 1209, 1346, 1488.

Observations *Gyroporus castaneus,* the only representative of this genus known from the state, is easily recognized by its small size, white tubes that become yellow as the spores mature, and the yellow spore deposit. In addition, the stipe and pileus are of the same pinkish-brown color. It is not likely to be confused with any other known bolete from California.

All collections cited above are made by Dr. Walter Sundberg, and this description has been drawn largely from his notes.

Although too small and rare in this state to be of any significance, in Europe, where it is often abundant, it is considered edible and choice.

LECCINUM S. F. GRAY

Nat. Arr. Brit. Pls. 1:640. Emended Snell, 1942

Pileus typically large, convex to pulvinate, becoming plano-convex to plane; surface dry to moist, less frequently subviscid or viscid, tomentose to velutinous to sometimes appressed fibrillose or rarely appressed fibrillose-scaly, white to whitish to gray or pink to orange to orange red to maroon; margin entire or with sterile "flaps" of cuticle tissue; context white, unchanging or becoming fuscous or blackish when exposed, in some passing through a reddish phase before darkening; taste and odor mild. Tubes 0.5–2 cm long, typically white to smoky gray or rarely pale yellow when young, becoming olive to dingy yellow with age; pores typically small, concolorous, often staining vinaceous or brown when bruised. Stipe stout, often clavate, solid; surface dry, conspicuously ornamented with small dots or squamules, ornamentations often white to tan when young, typically becoming brown or black with age, no annulus; context white, unchanging or changing as in pileus, sometimes changing to blue in base of stipe. Spores brown to dark brown in mass, ochraceous in KOH, often up to 17–20 μm long, subellipsoid to subfusoid, smooth. Hymenial cystidia scattered, often inconspicuous, hyaline, thin-walled, clavate to mucronate to fusoid-ventricose, usually not staining in KOH. Tube trama divergent. Pileus cuticle a tangled trichodermium, cells often disarticulating, pigment globules forming in some when mounted in Melzer's reagent. Stipe cuticle with numerous clusters of caulocystidia. Clamp connections absent.

Leccinum S. F. Gray

Key to Sections

1. Pileus margin with conspicuous segments or "flaps" of sterile tissue (check young basidiocarps)	Sect. *Leccinum*
1. Pileus margin without sterile segments or "flaps"	Sect. *Scabra*

Leccinum S. F. Gray

SECTION LECCINUM

Key to Species

1. Context of pileus and stipe unchanging or not becoming fuscous when exposed	*L. constans*	
1. Context eventually changing to fuscous when exposed		2
2. Context becoming red to reddish brown before changing to fuscous		3
2. Context changing directly to fuscous		5
3. Cells of cuticular hyphae developing pigment globules when mounted in Melzer's reagent	*L. aurantiacum*	
3. Pigment globules not developing in cells of cuticular hyphae when mounted in Melzer's reagent		4
4. Pileus pale brown to buff colored, viscid, changing to blue when bruised; context fading to white after darkening; found with madrone (*Arbutus*)	*L. arbuticola*	
4. Pileus brown to reddish brown to rust brown, dry to subviscid, context not changing to blue, remaining dark colored after exposure; found with aspens (*Populus*)	*L. discolor*	
5. Surface of pileus distinctly viscid		6
5. Surface of pileus moist to dry (sometimes subviscid when wet or old)		7
6. Pileus dark red; spores 4–5 μm broad	*L. manzanitae* var. *manzanitae*	
6. Pileus colored as above; spores narrowly elongate, 3–4 μm broad	*L. manzanitae* var. *angustisporum*	

7. Pileus dark dull brown to dark reddish brown; tubes and pores not becoming lavender when bruised	*L. brunneum*	
7. Basidiocarp not as above		8
8. Pileus pale buff to pale pink; stipe white or whitish during all stages	*L. armeniacum*	
8. Basidiocarp not as above		9
9. Pileus dark reddish brown; pores dark smoky olive; squamules on stipe small and densely crowded	*L. largentii*	
9. Basidiocarp not as above		10
10. Pileus brick red to ferruginous; surface tomentose to fibrillose when young	*L. insigne* var. *insigne*	
10. Pileus brown to reddish brown; surface glabrous when young	*L. insigne* var. *brunneum*	

LECCINUM ARBUTICOLA *sp. nov.*

Illustrations: See Microfiche No. 27

Pileus 7–10.5 cm latus, viscidus, glaber vel subtomentosus, brunneus vel ferrugineus, tactu caerulescens. Contextus albus, tactu testaceus demum fuscus vel nigrescens. Sapor et odor mitis. Tubuli usque 1.5 cm longi, brunnei demum bubalini; pori 0.5–1 mm lati, tactu olivaceo-brunnei. Stipes 9–13 cm longus, 1–3.5 cm crassus, aequalis, solidus, albus, squamis albis demum fuscis. Contextus stipitis albus, tactu apice pileo similis et basi caerulescens. Sporae 15–17.1 X 4.5–5.8 μm, ellipsoideae vel subcylindricae. Cystidia 41–46 X 9–11 μm, hyalina, fusoideo-ventricosa, dispersa vel rara. Cuticula intertexta, cellulis elongatis, incrustatis, globulis pigmenti nullis. Caulocystidia fasciculata, clavata, brunnea in KOH. Holotypus (no. 30514) a H. D. Thiers lectus prope Nevada City, Nevada County, November 11, 1972; in Herbarium San Francisco State University conservatus.

Pileus 7–10.5 cm broad, obtusely convex to convex when young, typically becoming plano-convex to occasionally convex with an incurved or decurved margin with age, occasionally with shallow depressions on the disc; surface viscid (at least in wet weather), viscidity much more apparent in older basidiocarps and often only slightly viscid when young, glabrous to obscurely tomentose, no evidence of appressed fibrils or scales (as in *L. manzanitae*), staining blue when bruised, especially when young; color brown to rust brown ("cinnamon brown" to "ochraceous tawny" to "Dresden brown" to "buckthorn brown"), little indication of reddish pigments, evenly colored or somewhat paler toward the margin, with age fading to pale brown ("buckthorn brown") to tan ("warm buff" to "light buff"), when covered with leaves or debris tan to buff ("clay color" to "warm buff" to "light buff") with many paler areas ("pale olive buff") or almost white, occasionally with darker areas ("cinnamon buff" to "pale ochraceous buff"); margin sterile. *Context* 2–3 cm thick, white, changing to reddish before becoming fuscous, reddish stage often persisting and readily noticeable; eventually the fuscous color fades and context becomes pallid. Taste and odor not distinctive.

Tubes up to 1.5 cm long, adnate to depressed when young, becoming deeply and narrowly depressed with age; when young dark brown ("Dresden brown"), with age fading to buff ("pale olive buff" to "olive buff"), staining darker olive buff to lavender ("avellaneous" to "wood brown") when bruised; *pores* 0.5–1 mm broad, staining olive brown when bruised.

Stipe 9–13 cm long, 1–3.5 cm broad at the apex; typically equal, but sometimes enlarging toward the base, solid, whitish mycelium at the base; surface dry, with scattered scabrosities, which never form a conspicuous element on the surface; background color white, but with a strong tendency to stain blue when bruised, color change immediate and suggestive of that of *Gyroporus cyanescens;* ornamentations white to pallid at first, typically becoming dark brown to fuscous, if covered with debris remaining pallid or white. *Context* white, staining as in pileus.

Spore print color unknown. *Spores* 15–17.1 X 4.5–5.8 μm, pale ochraceous to rarely dextrinoid in Melzer's, ochraceous in KOH, fusiform to subcylindric in face view, inequilateral in profile, walls smooth, thin.

Basidia 25–29 X 9–12 μm, one- to four-spored, clavate, hyaline, often filled with conspicuous oil droplets. *Hymenial cystidia* 41–46 X 9–11 μm, rare to scattered, embedded in the hymenium, often inconspicuous, hyaline, thin-walled, fusoid to fusoid-ventricose, often with an elongated and somewhat flexuous apex.

Tube trama hyaline, divergent from a distinct central strand, hyphae 5–6 μm wide. *Pileus trama* interwoven, hyaline in KOH, with scattered conspicuous laticifers, hyphae 5–7 μm wide. *Pileus cuticle* differentiated as a tangled trichodermium; orange yellow in ammonia, pale yellow in KOH and ochraceous in Melzer's; homogeneous, no pigment globules in Melzer's, some disarticulation, hyphae 8–12 μm wide, terminal cells elongated and with a tapered apex. *Caulocystidia* 25–36 X 7–10 μm, clustered, clavate to rarely fusoid-ventricose, contents when older staining brown in KOH, some secondarily septate, occasionally fertile basidia and huge caulocystidia intermixed. *Clamp connections* absent.

Chemical reactions KOH—context of stipe yellow; NH_4OH—context of pileus yellow.

Habit, habitat, and distribution Scattered to gregarious under madrone (*Arbutus menziesii*), often completely covered with leaves and debris. Known only from lower elevations of the Sierra Nevada in the vicinity of Grass Valley and Nevada City.

Material studied Nevada County: Thiers 30512, 30513, 30514 type, 30519, 30520.

Observations This is another addition to the ever-increasing number of Leccinums associated with members of the Ericaceae (manzanita, madrone, etc.). It is readily distinguished from the other species associated with these trees by the brown to buff color of the pileus, the distinct reddish stage in the sequence of color changes in the flesh with the eventual return of the context to almost white after prolonged exposure, and the strong blue discoloration developing in the basidiocarp when bruised. The absence of pigment globules in the cuticular hyphae in Melzer's reagent further differentiates it from *L. manzanitae.*

Edibility undetermined.

LECCINUM ARMENIACUM *Thiers, Mycologia 63:268. 1971*

Pileus 8–11 cm broad, convex to obtusely convex when young, becoming broadly convex to plano-convex to highly irregular at maturity; surface often shallowly and irregularly pitted, viscid to subviscid, glabrous to obscurely tomentose to fibrillose; color near apricot to pink ("apricot orange" to pale "orange-buff"), not becoming darker with age; margin with conspicuous sterile cuticular flaps. *Context* 1–2.5 cm thick, white when first exposed, slowly and irregularly changing to pale fuscous in some areas, not changing in all parts. Taste and odor not distinctive.

Tubes 1–2 cm long, depressed, whitish to pale olive ("pale olive-buff") during all stages, staining pale lilac ("avellaneous") to flesh color ("testaceous") when bruised; *pores* up to 1 mm, angular, concolorous.

Stipe 8–12 cm long, 1.5–3 cm thick at apex, clavate, solid; surface dry, squamulose, ground color white, squamules white to pallid, never becoming strongly discolored with age. *Context* white, changing slowly to fuscous in apical portion, slightly bluish discoloration in base.

Spore print brown. *Spores* 11–14.5 X 3.8–5 μm, subfusoid to subellipsoid, inequilateral, pale buff in KOH, pale ochraceous in Melzer's, walls smooth, thin.

Basidia 28–33 X 8–12 μm, four-spored, hyaline, clavate. *Hymenial cystidia* 27–40 X 7–15 μm, scattered to abundant, more numerous toward the pores, hyaline in KOH, fusoid to fusoid-ventricose with elongated, tapering apices, thin-walled.

Tube trama divergent from obscure central strand, hyaline in KOH, hyphae 5–7 μm wide. *Pileus trama* interwoven, homogeneous, hyphae 6–10 μm wide. *Pileus cuticle* differentiated as a trichodermium of tangled hyphal tips; hyaline in KOH, pale orange yellow in Melzer's; hyphae 8–12 μm wide, no disarticulation of cells, pigment globules not formed when mounted in Melzer's, terminal cells only slightly differentiated. *Caulocystidia* 30–45 X 10–15 μm, hyaline, clavate to subfusoid, rare large cells interspersed, numerous fertile basidia present. *Clamp connections* absent.

Chemical reactions KOH—context yellow in base of stipe; guaiac—positive on surface of pileus and base of stipe.

Habit, habitat, and distribution Solitary to gregarious in soil under madrone. It is known from San Mateo and Santa Cruz counties.

Material studied San Mateo County: Thiers 18193, 26983. Santa Cruz County: Thiers 12034.

Observations This species is easily recognized by the pale-pink to apricot color of the pileus and the white scales on the stipe, which never become dark brown or black. *Leccinum clavatum* appears somewhat similar but is readily distinguished by its darker and more intensely colored pileus, smaller spores, the pale-lilac color of the pores when bruised, and a different mycorrhizal host.

Edibility unknown.

LECCINUM AURANTIACUM *(Bulliard ex Saint-Amans) S. F. Gray, Nat. Arr. Brit. Pls. 1:646. 1821*

Boletus aurantiacus Persoon, Mycol. Europ. 2:147. 1825.

Ceriomyces viscidus Murrill, North Am. Flora, 9:139. 1910.

Leccinum versipelle Snell apud Slipp and Snell, Lloydia 7:58. 1944.

Illustrations: See Microfiche No. 28

Lange, J. E., and M. Lange, 600 Pilze in Farben, p. 189.

Leclair, A., and H. Essette, Les Bolets, pl. 60.

Romagnesi, H., Nouvel Atlas des Champignons, pl. 136.

Singer, R., Die Rohrlinge, Teil II, pl. XXIV, figs. 1–5.

Smith, A. H., Mushrooms in Their Natural Habitats, Reel V.

Smith, A. H., and H. D. Thiers, The Boletes of Michigan, pl. 70.

Snell, W. H., and E. A. Dick, The Boleti of Northeastern North America, pls. 68, 69.

Pileus 8–15 cm broad, convex to obtuse to pulvinate when young, becoming convex, broadly convex to plano-convex with age; surface dry, dull, uneven, or roughened, finely to matted-fibrillose when young, unchanging or appearing glabrous when older, sometimes finely and obscurely areolate; color bright rust brown to brick red ("ferruginous" to "rufous" to "apricot-orange") during all stages of development; margin incurved, with a conspicuous sterile band. *Context* 2–3 cm thick, white when first exposed, changing to vinaceous to near flesh color ("testaceous"), then to gray to bluish gray ("fuscous") or black. Taste and odor mild.

Tubes 1–2 cm long, adnate to depressed when young, becoming deeply and narrowly depressed with age; when young pale olive ("pale olive-buff" to "deep olive-buff" to "olive-buff"), darkening to olive drab ("dark olive-buff") or becoming brown ("Wood brown") at maturity; *pores* 0.5–1 mm broad, angular, staining olive to olive brown to vinaceous brown when bruised.

Stipe 10–17 cm long, 2–3 cm thick at apex, clavate to ventricose, becoming equal, often narrowed at the base and apex, solid; surface dry, ground color whitish to white, occasionally staining blue in base, squamules pallid at first, black or dark brown with age. *Context* white, staining as in pileus.

Spore print brown. *Spores* 12–13.5 X 3.5–4.5 μm (spores from single-spored basidia up to 28 μm long), fusoid to subellipsoid, pale ochraceous in KOH, yellow brown in Melzer's, walls smooth, thin.

Basidia 30–55 X 9–11 μm, clavate, hyaline in KOH, one- to four-spored. *Hymenial cystidia* 30–40 X 9–11 μm, clavate to fusoid, scattered, deeply embedded, thin-walled, hyaline in KOH.

Tube trama hyaline in KOH, divergent, hyphae 5–8 μm wide. *Pileus trama* interwoven, hyaline in KOH, homogeneous, hyphae 4–7 μm wide. *Pileus cuticle* differentiated as a tangled trichodermium, pale ochraceous in KOH, hyphae 7–11 μm wide, considerable disarticulation of cells noted, walls often asperulate, pigment globules formed when mounted in Melzer's reagent, terminal cells not strongly differentiated, no short intermediate cells present. *Caulocystidia* 38–48 X 8–13 μm, dark brown in KOH, clavate, with large fusoid-ventricose cells interspersed. *Clamp connections* absent.

Chemical reactions $FeSO_4$—context blue gray.

Habit, habitat, and distribution Scattered to gregarious in soil under pines in a forest composed largely of pines and aspens.

Material studied Madera County: Thiers 23608, 23611.

Observations It has been common practice to identify every rust-red-colored *Leccinum* as this species; however, a critical reevaluation of this group has shown it to be composed of numerous species that can be easily distinguished. *Leccinum aurantiacum* sensu strictu is characterized by the brick-red-colored pileus, flesh that turns reddish before becoming bluish black, and the development of pigment globules in cells of the cuticular hyphae when they are mounted in Melzer's reagent. As can be seen from the list of material studied, this is actually a rather rare species in California and has only been found in the vicinity of the Devil's Post Pile National Monument.

Edible and generally considered good. The dark discoloration of the flesh when dried or cooked is objectionable to some.

LECCINUM BRUNNEUM *Thiers, Mycologia 63:269. 1971*

Pileus 6–10 cm broad, convex when young, becoming broadly convex to pulvinate to plano-convex to irregular in outline at maturity; surface dry, but may be viscid or subviscid in wet weather, dull, glabrous to finely subtomentose or obscurely appressed fibrillose, occasionally becoming somewhat fibrillose with age; color dark brown ("bister" to "warm sepia" to "Prout's brown"), sometimes somewhat paler brown ("cinnamon-brown" to "Verona brown" to "snuff brown") with age, drying dark dull brown; margin incurved, with conspicuous sterile cuticular flaps. *Context* 1–2.5 cm thick, white when first exposed, changing to fuscous with no reddish intermediate phase, fuscous color sometimes fading after prolonged exposure. Taste and odor mild.

Tubes 1–2 cm in length, depressed, white when young, becoming pale olive ("pale olive-buff") when mature, staining brown ("clay color" to "buckthorn brown" to "Mikado brown") when bruised, color fading in some after bruising; *pores* up to 1 mm broad, angular, concolorous.

Stipe 8–15 cm long, 1.5–3(4) cm thick at the apex, clavate, solid; surface dry, covered with small squamules, ground color white, squamules white at first, becoming darker with maturity. *Context* white when first exposed, then becoming fuscous directly as in the pileus, sometimes becoming blue in the base.

Spore print brown. *Spores* (14)15.5–19.5 X 4–6 μm, ochraceous in KOH, some showing a dextrinoid reaction in Melzer's, fusoid to subcylindric to subellipsoid, inequilateral, walls smooth, moderately thick.

Basidia 30–35 X 7–11 μm, hyaline in KOH, clavate to pyriform, four-spored. *Hymenial cystidia* 35–50 X 8–12 μm, scattered to abundant, often inconspicuous; some staining pale brown in KOH and dark reddish brown in Melzer's, fusoid to fusoid-ventricose to clavate.

Tube trama hyaline, divergent from a distinct central strand, hyphae 3–7 μm wide. *Pileus trama* interwoven, homogeneous, hyaline in KOH, hyphae 5–9 μm wide. *Pileus cuticle* differentiated as a trichodermium of tangled hyphal tips; pale ochraceous in KOH, tawny to russet in Melzer's; hyphae 11–15 μm wide, walls sometimes asperulate, pigment globules not forming when mounted in Melzer's, terminal cells relatively long and narrow, tapered somewhat at the tip. *Caulocystidia* 24–37 X 7–12 μm, clavate to fusoid to clavate with elongated, tapering apices, typically dark brown, but sometimes hyaline in KOH. *Clamp connections* absent.

Chemical reactions No distinctive reactions observed.

Habit, habitat, and distribution Gregarious to scattered in soil under aspens. Known only from the higher elevations in the Sierra Nevada, perhaps due to a similar distributional restriction of its only known mycorrhizal host.

Material studied Alpine County: Thiers 20818. El Dorado County: Thiers 20745, 20747, 20749, 20751, 20753, 20756, 20759, 20767. Mono County: Thiers 20963.

Observations The dull-dark-brown color of the pileus readily distinguishes this species. It is most likely to be confused with *L. fibrillosum,* but in *L. brunneum* the color changes of the exposed context are different and the pileus is not strongly fibrillose. The difference in the color of the cap and the lack of incrustations on the cuticular hyphae of *L. brunneum* distinguish it from *L. obscurum. Leccinum subspadiceum,* another brown species, is distinct because of the smaller spores, cinnamon-buff pores, and the yellow color in the dried stipe.

Edibility unknown.

LECCINUM CONSTANS *sp. nov.*

Illustrations: See Microfiche No. 29

Pileus 10–14 cm latus, uvidus vel subviscidus, glaber in centro, appresso-fibrillosus ad marginem, subpuniceus, fibrillulis parce furvioribus. Contextus 2.5 cm crassus, albus, immutabilis. Tubuli 2 cm longi, albi demum subolivacci; pori 1 mm lati, concolores tactu brunnei. Stipes 12–15 cm longus, 2–3 cm crassus, siccus, fibrilloso-squamosus, albus, squamis fuscis; contextus albus, immutabilis. Sporae 15.5–18 X 4–5 μm, fusoideae vel subellipsoideae. Cystidia 24–31 X 7–9 μm, inconspicua, fusoidea vel submucronata. Cuticula intertexta, globulis pigmenti nullis (in Melzer's). Holotypus (no. 26971) a H. D. Thiers lectus prope Boulder Creek, Santa Cruz County, December 6, 1970; in Herbarium San Francisco State University conservatus.

Pileus 10–14 cm broad when mature, convex to pulvinate during all stages of development; surface moist to subviscid, more or less glabrous on the disc, becoming closely appressed fibrillose toward the margin; ground color pale pink ("pinkish buff" to "pale pinkish buff"), fibrils somewhat darker ("apricot buff" to "flesh ocher" to "ochraceous-salmon"); margin with sterile flaps of cuticular tissue. *Context* up to 2.5 cm thick, white, unchanging when exposed.

Tubes up to 2 cm in length, white to whitish when young, changing to pale olive ("olive-buff") when bruised; *pores* less than 1 mm broad, angular, concolorous, staining brown when bruised.

Stipe 12–15 cm long, 2–3 cm thick at the apex, clavate to subclavate, solid; surface dry, densely squamulose or furfuraceous, ground color white, squamules whitish to pale brown when young, becoming dark brown with maturity. *Context* white, unchanging in any portion when exposed.

Spore print unknown. *Spores* 15.5–18 X 4–5 μm, fusoid to subellipsoid in face view, inequilateral in profile, pale ochraceous in Melzer's and KOH, walls smooth, moderately thick.

Basidia 25–31 X 6–9 μm, hyaline in KOH, four-spored, occasionally one- to two-spored, clavate. *Hymenial cystidia* 24–31 X 7–9 μm rare, inconspicuous, hyaline, thin-walled, fusoid to submucronate.

Tube trama hyaline, divergent, hyphae 5–6 μm wide. *Pileus trama* hyaline, interwoven, homogeneous. *Pileus cuticle* differentiated as a tangled trichodermium of septate hyphal tips, considerable disarticulation of cells, terminal cell not differentiated, cell walls sometimes asperulate, pigment globules not formed in Melzer's. Scales on stipe composed of clavate caulocystidia that stain brown in KOH. *Clamp connections* absent.

Chemical reactions unknown.

Habit, habitat, and distribution Solitary in soil in dense hardwood coastal forests near Santa Cruz.

Material studied Santa Cruz County: Thiers 26971 type.

Observations This is another species of *Leccinum* apparently associated with madrone. Although similar in gross appearance to *L. manzanitae*, *L. constans* is easily distinguished by the absence of a viscid pileus, the noticeably paler color of the pileus, and, most significantly, the absence of any color change in the context of either the pileus or stipe. This species belongs to the subsection *Vulpina* of the section *Leccinum*, but it is difficult to predict its affinities within that section.

Edibility undetermined.

LECCINUM DISCOLOR *Smith, Thiers and Watling, Mich. Botan. 5:152. 1966*

Illustrations: See Microfiche No. 30

Pileus 10–15 cm broad, globose to convex when young, becoming broadly convex to plano-convex to depressed on the disc to occasionally highly irregular in outline; surface dry to subviscid, dull, often shallowly pitted or reticulate, glabrous but often appearing tomentose or fibrillose, sometimes rimose-areolate near the margin; color brown ("Kaiser brown" to "cinnamon-brown") to orange red ("russet" to "auburn" to "rufous") to sometimes pale buff ("pale pinkish buff" to "cinnamon-buff"); margin with typical sterile flaps of cuticular tissue. *Context* 1–3 cm thick, white when first exposed, changing to flesh color ("testaceous") to pink ("pinkish cinnamon") then blackish ("fuscous"). Taste and odor mild.

Tubes 1–2 cm in length, depressed, white when young, becoming pale olive brown ("pale olive-buff" to "olive-buff") at maturity, staining dark brown when bruised; *pores* angular, 0.5–1 mm broad, concolorous.

Stipe 6–12 cm long, 1–3 cm thick at apex, equal to clavate, white mycelium at the base, solid; surface dry, conspicuously ornamented, ground color white to whitish, squamules white to pallid when young, becoming dark brown to black with age, often somewhat crowded when young, becoming separated at maturity. *Context* white, undergoing color changes similar to pileus, bluing not observed.

Spore print brown. *Spores* 15.5–20 X 4.5–6.5 μm, fusoid to subcylindric, inequilateral, ochraceous in KOH and Melzer's, some showing a dextrinoid reaction, walls smooth, thin.

Basidia 30–37 X 9–12 μm, clavate to pyriform, hyaline and with granular contents in KOH, four-spored. *Hymenial cystidia* 30–46 X 8–15 μm, hyaline to sometimes appearing filled with dark brown amorphous material in KOH, fusoid to clavate with elongated, tapering apices, thin-walled.

Tube trama divergent from a distinct to poorly developed central strand, hyaline to pale ochraceous in KOH, hyphae 3–4 μm wide, often appearing gelatinous. *Pileus trama* interwoven, homogeneous, hyaline in KOH, hyphae 8–16 μm wide. *Pileus cuticle* differentiated as a tangled trichodermium of free hyphal tips; buff colored in KOH, ochraceous tawny in Melzer's; hyphae 8–10 μm wide, walls smooth to asperulate and sometimes appearing incrusted, considerable disarticulation of cells noted, terminal cells only slightly narrowed at apex. *Caulocystidia* 25–45 X 10–17 μm, clavate to mucronate to subfusoid, mostly dark brown in KOH but some hyaline cells present, an occasional large cell interspersed, walls smooth, thin. *Clamp connections* absent.

Chemical reactions unknown.

Habit, habitat, and distribution Solitary to scattered in soil under aspens. Known only from Luther Pass in the Sierra Nevada.

Material studied El Dorado County: Thiers 20744, 20750, 20755, 20766, 20768, 20769, 20770, 23564, 23565, 23566, 23626, 23627.

Observations This is one of the species previously identified as *L. aurantiacum*. The most obvious characteristic distinguishing the two species is found in the appearance of the cuticular hyphae in Melzer's. The true *L. aurantiacum* has an abundance of pigment globules formed within the cells, whereas in *L. discolor* the contents remain homogeneous or only become finely granular. The absence of any reddish phase in the discoloration of the exposed context distinguishes *L. insigne* from *L. discolor*.

Edible.

LECCINUM INSIGNE *var.* INSIGNE *Smith, Thiers, and Watling, Mich. Botan. 5:160. 1966*

Illustrations: See Microfiche No. 31
Smith, A. H., and H. D. Thiers, The Boletes of Michigan, pl. 77

Pileus 6–17 cm broad, convex to globose when young, becoming broadly convex to plano-convex to highly irregular with broad, shallow depressions or pits at maturity; surface dry to moist, occasionally subviscid in wet weather, glabrous to finely velutinous to subtomentose when young, unchanging or sometimes becoming obscurely appressed-fibrillose or fibrillose-scaly with age, frequently finely areolate to rimose-areolate when old; color somewhat variable, but typically some shade of reddish brown ("cinnamon" to "tawny" to "cinnamon-rufous" to "hazel" to "ferruginous" to as dark as "Kaiser brown" to as pale as "pinkish cinnamon"), often somewhat variegated with age; margin incurved, with noticeable sterile cuticular flaps. *Context* 1.5–3 cm thick, white when first exposed, changing to fuscous with no reddish intermediate stage, color changes sometimes very slow in developing. Taste and odor mild.

Tubes 1–2 cm long, narrowly depressed at first, becoming broadly and deeply depressed with age, somewhat ventricose in arrangement, white when young, olive buff ("pale olive-buff") to yellowish ("cartridge buff" to "chamois") when mature, staining brown when bruised; *pores* up to 1 mm broad, angular, concolorous.

Stipe 8–13 cm long, 1–2 cm thick at the apex, equal to clavate, solid; surface dry, squamulose, ground color white to whitish, squamules close and pallid when young, typically becoming dark brown to black and often more widely spaced at maturity, often very fine, and sometimes more or less collapsing with age. *Context* white, changing to fuscous as in pileus, blue discoloration sometimes developing in the base.

Spore print brown. *Spores* (11)13–16(18) X 4–6 μm, subfusoid to subcylindric, inequilateral, pale ochraceous in KOH, ochraceous in Melzer's, walls smooth, moderately thickened.

Basidia 29–34 X 8–12 μm, hyaline in KOH, clavate to pyriform, four-spored. *Hymenial cystidia* 27–43 X 7–13 μm, clavate to clavate with elongated, tapering apices, scattered to numerous, hyaline to dark brown in KOH.

Tube trama divergent from a distinct central strand, hyphae 3–5 μm wide. *Pileus trama* homogeneous, interwoven, hyphae 8–12 μm wide. *Pileus cuticle* differentiated as a trichodermium of tangled hyphal tips, hyphae 4–7 μm wide, walls sometimes roughened, pigment globules not developing when mounted in Melzer's, terminal cells with obtuse apices, or with tapered apices. *Caulocystidia* 29–39 X 10–14 μm, clavate to fusoid, dark brown in KOH, thin-walled. *Clamp connections* absent.

Chemical reactions No distinctive reactions observed.

Habit, habitat, and distribution Gregarious to scattered in soil under aspens. Known only from the higher elevations in the sierra Nevada and Cascades.

Material studied El Dorado County: Thiers 20746, 20748, 23625. Madera County: Thiers 23612. Nevada County: Thiers 13164. Sierra County: Thiers 13238, 21146, 23597, 23630, 23946.

Observations This species is probably the most common and widely distributed *Leccinum* in North America and was formerly included in the broad concept of *L. aurantiacum* and *L. discolor. Leccinum insigne* is readily distinguished by the characteristic color changes occurring in the flesh when exposed, the rust-red color of the pileus, and the absence of pigment globules in the cuticular hyphae when mounted in Melzer's. Throughout its known range it appears to be associated with aspens, which most probably accounts for its absence in the coastal forests of this state.

Edible.

LECCINUM INSIGNE *var.* BRUNNEUM *Thiers, Mycologia 63:272. 1971*

Illustrations: See Microfiche No. 32

Pileus 10–15 cm broad, convex, becoming broadly convex to plano-convex; surface dry, dull, glabrous to obscurely velutinous on the margin when young, unchanging or becoming finely and obscurely appressed-fibrillose when older, occasionally finely areolate in some areas with age; color when young brown ("Mars brown" to "Prout's brown" to "walnut brown" to "Rood's brown"), fading to reddish brown ("hazel" to "tawny" to "Kaiser brown" to "russet") at maturity; margin incurved, with conspicuous sterile cuticular flaps. *Context* 1.5–3 cm thick, white when first exposed, changing to fuscous with no intermediate reddish stage. Taste and odor mild.

Tubes 1–3 cm long, deeply depressed, pale olive ("pale olive-buff"), staining vinaceous to purple, then eventually becoming fuscous when bruised, not becoming brown; *pores* up to 1 mm broad, often slightly darker olive ("olive-buff" to "dark olive-buff") than the tubes, when older with numerous dark olive ("olive-green") spots.

Stipe 10–15 cm long, 2–2.5 cm thick at the apex, clavate, solid; surface dry, squamulose, ground color white or whitish, squamules pallid when young, becoming black and conspicuous with age. *Context* white, undergoing similar color changes as in the pileus, not changing to blue.

Spore print brown. *Spores* 14–16.5 X 5–6.5 μm, subfusoid to subcylindric, inequilateral, pale ochraceous in KOH, ochraceous in Melzer's, walls smooth, thin.

Basidia 30–35 X 8–10 μm, subclavate to clavate, four-spored, hyaline. *Hymenial cystidia* 40–60 X 8–15 μm, hyaline in KOH, common, inconspicuous, fusoid-ventricose, thin-walled.

Tube trama divergent, hyaline in KOH, hyphae 4–7 μm wide. *Pileus trama* interwoven, homogeneous, hyphae 11–15 μm wide. *Pileus cuticle* differentiated as a trichodermium of tangled hyphal tips, cells somewhat sausage-shaped, disarticulating, walls smooth to finely asperulate, pigment globules not formed in Melzer's, terminal cells not strongly differentiated, up to 14 μm broad. *Caulocystidia* 25–40 X 9–15 μm, clavate, dark brown in KOH, some large cells interspersed. *Clamp connections* absent.

Chemical reactions No distinctive reactions observed.

Habit, habitat, and distribution Scattered to gregarious in humus under aspens. Known only from near Devil's Post Pile National Monument.

Material studied Madera County: Thiers 23609, 23610.

Observations The dark-brown pileus when young with a tendency to fade somewhat with age and the dull, dark-brown color when dried readily distinguish this variety from others that have been described.

Edibility undetermined.

LECCINUM LARGENTII *sp. nov.*

Illustrations: See Microfiche No. 33

Pileus 8–15 cm latus, convexus vel plano-convexus, siccus interdum viscidus aetate, appresso-fibrillosus vel fibrilloso-squamosus margine, disco tomentosus vel velutinus, obscure rufus. Contextus albus tactu fuscus. Tubuli 1–2 cm longi, fusco-olivacei; pori fumoso-brunnei. Stipes 10–12 cm longus, 1.5–3.5 cm latus, siccus, albus, squamis confertis, fuscis, parvulis. Sporae 13.3–17.1 X 4–6 μm, subcylindricae vel subellipsoideae. Cuticula intertexta, globulis pigmenti (in Melzer's). Holotypus (no. 28531) a David Largent lectus prope Eureka, Humboldt County, October 30, 1971; in Herbarium San Francisco State University conservatus.

Pileus 8–15 cm broad, convex when young, becoming broadly convex to pulvinate to plano-convex to somewhat irregular at maturity; surface dry, becoming subviscid to viscid only when wet or very old, appressed fibrillose to appressed fibrillose-scaly, sometimes conspicuously so near the margin, innately fibrillose to tomentose to velutinous on the disc, sometimes appearing somewhat glabrous on the disc with age, some often obscurely and shallowly areolate when very old; color deep, dark reddish brown ("bay" to "mahogany red" to "claret brown") to pale reddish brown ("chestnut" to "auburn"), near the margin sometimes slightly paler—red to orange red ("apricot orange" to "scarlet"); margin incurved, with very dark colored sterile flaps or sterile margin entire and not becoming broken into flaps. *Context* 1–3 cm thick, white, slowly and erratically becoming directly fuscous when exposed, most conspicuous and often only change occurring at the juncture of stipe and pileus, change much more pronounced than in *L. manzanitae.* Taste and odor mild.

Tubes 1–2 cm long, shallowly depressed; color pale olive ("olive-buff" to "pale olive-buff"); *pores* small, less than 1 mm broad, angular in outline, smoke colored ("dark olive-buff" to "citrine-drab"), unchanging or becoming dark olivaceous ("Saccardo's umber" to "deep olive") with age, staining deep olive when bruised.

Stipe 10–12 cm long, 1.5–3.5 cm thick, ventricose to tapering toward the base, solid; surface dry, covered with very fine, densely packed squamules colored pale brown ("clay color"), with age becoming dark brown to black, but remaining small and dense; in some, the surface becomes reticulate to almost alveolate from the elaboration of the squamules; background color white during all stages. *Context* white, changing to fuscous at the apex, often with a reddish or bluish discoloration at the base.

Spore print color not determined. *Spores* 13.3–17.1 X 4–6 μm, pale ochraceous in KOH and Melzer's, subcylindric to subellipsoid, inequilateral, walls smooth, moderately thick.

Basidia 26–34 X 7–10 μm, clavate to pyriform, hyaline in KOH and Melzer's, four-spored. *Hymenial cystidia* 21–28 X 5–7 μm, scattered to rare, embedded, hyaline, very thin-walled, inconspicuous, fusoid to somewhat fusoid-ventricose with an elongated, narroed apex.

Tube trama slightly divergent from a central strand, hyaline in KOH, hyphae 5–7 μm wide. *Pileus trama* hyaline in KOH, interwoven, homogeneous, hyphae 6–9 μm wide. *Pileus cuticle* differentiated as a tangled trichodermium, terminal cells tapered noticeably at the apex, considerable disarticulation of the hyphal cells, hyphae 7–9 μm wide, walls smooth, contents ochraceous to cinnamon in KOH, pigment globules developing when mounted in Melzer's. *Caulocystidia* 40–55 X 12–20 μm, clavate, thin-walled, cinnamon to ochraceous in KOH, bright rust ochraceous in Melzer's. *Clamp connections* absent.

Chemical reactions $FeSO_4$—context pale gray, pores very dark olive; guiaic—weakly positive on context in base of stipe.

Habit, habitat, and distribution Gregarious to scattered under toyon (*Heteromeles arbutifolia*). Known only from the vicinity of Samoa near Eureka.

Material studied Humboldt County: Thiers 28531 type. (Collection made by David Largent.)

Observations This is another of the large, massive boletes found associated with hardwoods in the coastal forests of northern California and is apparently related to *L. manzanitae. Leccinum largentii,* however, has a dry pileus that is often noticeably appressed fibrillose-scaly on the margin, very dark smoky olive pores that do not change color with age, and a highly distinctive elaboration of the squamules on the stipe. The last character is perhaps the most distinctive of all. This species is somewhat suggestive of *L. fallax,* which occurs in the Rocky Mountains under spruce and fir, has narrower spores, and pallid squamules on the stipe.

Edibility unknown.

LECCINUM MANZANITAE *var.* MANZANITAE *Thiers, Mycologia 63:266. 1971*

Illustrations: See Microfiche No. 34

Pileus (5)8–20 cm broad, globose to convex when young, broadly convex to pulvinate when mature; surface often shallowly to deeply pitted or reticulate, viscid to subviscid, especially with age, strongly appressed-fibrillose during all stages of development or occasionally tomentose, never glabrous, fibrils often more conspicuous toward the margin; color dark red ("bay" to "mahogany red" to "Mars brown" to "burnt umber") during all stages of development; margin incurved, with conspicuous sterile projecting cuticular segments. *Context* 2–4 cm thick, white when first exposed, slowly and erratically changing to fuscous with no reddish intermediate stage; color changes often more pronounced in young basidiocarps. Taste and odor mild.

Tubes 1–2.5 cm in length, adnate to shallowly or deeply depressed, pale olive ("pale olive-buff") when young, darkening to olive drab ("olive-buff" to "deep olive-buff") at maturity, staining dark brown ("buffy brown") when bruised; *pores* up to 1 mm broad, angular, concolorous.

Stipe 10–16 cm long, 1.5–3.5 cm thick at the apex, clavate to ventricose, solid; surface dry, conspicuously squamulose, squamules typically pallid when young, darkening to near fuscous with age, ground color white to whitish. *Context* white, slowly staining fuscous in apical portion when exposed, sometimes bluing in the base.

Spore print brown. *Spores* 13–17 X 4–5.5 μm, ochraceous in KOH and Melzer's, fusoid to subellipsoid to subcylindric, inequilateral; walls smooth, moderately thick.

Basidia 27–32 X 6–9 μm, clavate to pyriform, hyaline in KOH, four-spored. *Hymenial cystidia* 23–32 X 4–6 μm, often obscure, scattered, hyaline in KOH, fusoid to clavate with narrow, elongated apices.

Tube trama obscurely divergent from a distinct central strand, hyaline in KOH, hyphae 6–8 μm wide. *Pileus trama* hyaline in KOH, interwoven, homogeneous, hyphae 6–9 μm wide. *Pileus cuticle* differentiated as a trichodermium of free, tangled hyphal tips, no evidence of gelatinization of hyphal walls, hyphae 8–12 μm wide, some disarticulation of cells noted, walls smooth to sometimes obscurely roughened, contents ochraceous in KOH, reddish in Melzer's, pigment globules forming when mounted in Melzer's, terminal cells elongated and often noticeably tapered. *Caulocystidia* 35–45 X 9–14 μm, clavate to mucronate to fusoid, staining dark brown in Melzer's and KOH, thin-walled, large cells occasionally interspersed, basidia sometimes present. *Clamp connections* absent.

Chemical reactions KOH—tubes pale red, then blackening; HNO_3—tubes orange yellow; $FeSO_4$—context pale gray.

Habit, habitat, and distribution Solitary to scattered in soil under madrone and manzanita. This is by far the most common *Leccinum* in the coastal forests of California, where it has been observed associated only with ericaceous hosts. It has not yet been found in the foothills or mountain ranges. In the Santa Cruz Peninsula, where it is often abundant, it is frequently referred to as the "manzanita mushroom."

Material studied Marin County: Thiers 24654. Mendocino County: Thiers 8299, 9283, 10671, 18152, 21412, 21413, 23064, 24189, 24450, 24461, 24462, 24507, 24737. San Mateo County: Thiers 11200, 18059, 18336, 24735, 24736, 24755. Santa Cruz County: Thiers 12034, 14404.

Observations This is one of the large, massive boletes with caps that often weigh several pounds. It is distinguished by the very dark red color of the pileus, which is characteristically viscid even in dry weather, the conspicuously fibrillose surface of the cap, and the darkening of the context of the cap and stipe apex when exposed. These color changes are often somewhat erratic and may be very slow in developing; frequently, there are only localized areas in which the color change can be detected. *Leccinum manzanitae* is somewhat suggestive of *L. ponderosum*, but the context of the latter does not darken when exposed, it is not as deeply colored, and the pileus is glabrous. *Leccinum constans*, also associated with madrone in the coastal regions, is much paler in color, and the context remains unchanged in color when exposed to the air.

This fungus is one of the common edible mushrooms, and it is often gathered and dried as is done with *Boletus edulis*.

LECCINUM MANZANITAE *var.* ANGUSTISPORUM *Thiers, Mycologia 63:267. 1971*

As in var. *manzanitae,* except that the spores reach only 3–4 μm in width and often average 1–2 μm longer. In all other aspects this variant is similar to the typical variety.

Habit, habitat, and distribution Solitary in soil under madrone and manzanita in coastal forests.

Material studied Mendocino County: Pando 11-9-69; Thiers 8774.

Edibility undetermined.

Leccinum S. F. Gray

SECTION SCABRA

Key to Species

1. Pileus gray to gray brown	*L. montanum*
1. Pileus whitish to pale tan or tan	*L. californicum*

LECCINUM CALIFORNICUM *Thiers, Mycologia 63:273. 1971*

Pileus 5–7.5 cm broad, globose to convex when young, becoming plane to plano-convex when mature; surface smooth, glabrous when young, somewhat fibrillose and often strongly rimose-areolate with age; color almost white to pale buff ("pale pinkish buff" to "pale olive-buff") with age, sometimes darkening to buff color ("pinkish buff" to "vinaceous-buff"); margin entire, lacking sterile cuticular flaps. *Context* 1–2 cm thick, white, unchanging when exposed. Taste and odor mild.

Tubes 1–1.5 cm in length, depressed, white to pale buff ("pale olive-buff") when young, darkening to olive brown ("olive-buff") and eventually becoming lavender brown ("avellaneous" to "vinaceous-buff" to "wood brown"), becoming dark brown when bruised; *pores* up to 1 mm broad, angular, concolorous.

Stipe 5–8 cm long, 1–1.5 cm thick at the apex, equal to subclavate; surface dry, grayish yellow in basal region, otherwise white to whitish, squamulose, squamules white to pallid, remaining so or darkening somewhat when very old; solid. *Context* white, unchanging or becoming yellowish in the base.

Spore print brown. *Spores* 14.5–17.5 X 5.5–6.5 µm, dark ochraceous in KOH, rarely dextrinoid in Melzer's, fusoid to subcylindric, inequilateral, walls smooth, moderately thick.

Basidia 32–37 X 9–12 µm, hyaline in KOH, four-spored, clavate to pyriform. *Hymenial cystidia* 45–55 X 7–10 µm, inconspicuous, scattered to rare, fusoid to clavate with elongated, tapering apices, hyaline, very thin-walled.

Tube trama obscurely divergent, hyaline in KOH, hyphae 4–6 µm wide. *Pileus trama* interwoven, homogeneous, hyphae 8–12 µm wide. *Pileus cuticle* differentiated as a tangled trichodermium of free hyphal tips, bright red in Melzer's, almost hyaline in KOH, hyphae 7–10 µm wide, some disarticulation noted, no pigment globules developing when mounted in Melzer's, terminal cells slightly narrowed. *Caulocystidia* 40–55 X 10–17 µm, bright red ("russet") in Melzer's, clavate to mucronate. *Clamp connections* absent.

Chemical reactions No distinctive reactions observed.

Habit, habitat, and distribution Gregarious in soil under aspens at Grass Lake, Luther Pass, El Dorado County. Known only from the type locality.

Material studied El Dorado County: Thiers 23624.

Observations This species can be distinguished from *L. holopus,* to which it appears most similar, by the reddish discoloration of the caulocystidia when mounted in Melzer's and the considerably larger cystidia. Also, *L. holopus* has a greenish cast in the pileus and the flesh changes color when exposed, but *L. californicum* shows neither of these characteristics. The white to pallid color of the pileus of *L. californicum* readily distinguishes it from *L. montanum,* which occurs in the same area.

Edibility not determined.

LECCINUM MONTANUM *Thiers, Mycologia 63:274. 1971*

Illustrations: See Microfiche No. 35

Pileus 3–8 cm broad, convex to broadly convex when young, becoming broadly convex to plano-convex to shallowly depressed with an uplifted margin at maturity; surface dry to sometimes moist but not viscid, glabrous to occasionally somewhat tomentose or appressed fibrillose, often shallowly and irregularly pitted or roughened; color brown to gray brown ("buffy brown" to "drab" to "cinnamon-brown"), somewhat paler ("cinnamon-buff" to "pinkish cinnamon") toward the margin when older; margin entire, without sterile cuticular flaps. *Context* 1–2 cm thick, white, unchanging or rarely erratically and irregularly becoming pale watery gray upon exposure. Taste and odor mild.

Tubes 1–2 cm in length, depressed, white to buff ("pale pinkish buff" to "pale olive-buff") when young, becoming pinkish brown ("light pinkish cinnamon"), and eventually becoming lavender brown ("avellaneous" to "wood brown") at maturity, staining yellow brown to dark brown when bruised; *pores* up to 1 mm broad, angular, concolorous.

Stipe 5–10 cm in length, 1–1.5 cm thick at the apex, equal, solid; surface dry, conspicuously squamulose, ground color white to whitish, squamules white to pallid when young, becoming brown in apical portion and black toward the base with age, often small and inconspicuous with age. *Context* white, unchanging when exposed or rarely developing scattered watery spots, rarely bluing in the base.

Spore print brown. *Spores* 14–17.5(20) X 4–6 μm, pale ochraceous in KOH, some dextrinoid in Melzer's, fusoid to subcylindric to subellipsoid, inequilateral, walls smooth, thick.

Basidia 24–35 X 8–10 μm, four-spored, hyaline in KOH, clavate to pyriform. *Hymenial cystidia* 30–50(63) X (7)12–16 μm, very obscure, hyaline, thin-walled, scattered to rare, clavate with elongated, tapering flexuous apices.

Tube trama divergent from a distinct central strand, hyaline except for pale buff color of strand, hyphae 4–7 μm wide. *Pileus trama* loosely interwoven, homogeneous, hyphae 7–10 μm wide. *Pileus cuticle* differentiated as a tangled trichodermium with numerous free hyphal tips, hyphae 5–8 μm wide, occasional large, somewhat fusoid-shaped cell interspersed among smaller cells, walls smooth, some tendency for pigment globules to form when mounted in Melzer's, terminal cells with apex slightly narrowed or tapered. *Caulocystidia* 30–50 X 8–15 μm, clustered, brown in KOH, clavate to pyriform to mucronate. *Clamp connections* absent.

Chemical reactions No distinctive reactions observed.

Habit, habitat, and distribution Scattered to gregarious in soil under aspens. Known only from the vicinity of Grass Lake at Luther Pass in the Sierra Nevada.

Material studied El Dorado County: Thiers 20757, 20828, 23562, 23563, 23638, 23629.

Observations *Leccinum montanum* is distinguished from *L. scabrum*, which it superficially resembles, by the presence of cystidia along the sides of the tubes, the absence of any pinkish discoloration in the flesh when exposed, the dry to moist but not viscid pileus, and a different mycorrhizal host. The discovery of this species and *L. californicum* mark the first report of any species belonging to this section of the genus from this state. *Leccinum californicum* is easily distinguished by the pallid to white color of the basidiocarp.

Edibility undetermined.

PULVEROBOLETUS MURRILL

Mycologia 1:9. 1909

Pileus convex to plano-convex becoming plane to plano-convex; surface, at least when young, covered with an intensely yellow veil that obscures the surface color, dry, pulverulent, moist to subviscid or viscid when veil removed, pinkish cinnamon to brown, often with reddish streaks or spots, frequently pink to reddish with age; margin incurved, often with conspicuous veil fragments attached; context white to pale yellow, slowly changing to blue when exposed; taste mild to acid, odor mild. Tubes 0.5–1.5 cm long, yellow during all stages, changing to blue when bruised; pores up to 1 mm broad, angular, concolorous. Stipe up to 11 cm long, equal to tapering slightly toward the apex; surface dry to subviscid, covered with bright yellow veil, at least when young. Spores olive brown in mass, fusoid to ovoid, smooth, thin-walled, pale ochraceous in KOH. Hymenial cystidia scattered to numerous, hyaline to yellow in KOH, clavate to fusoid to subcylindric. Tube trama divergent. Cuticle loosely interwoven, no sphaerocysts in veil. Clamp connections absent.

Pulveroboletus Murrill

Key to Species

Only one species known from California *P. ravenelii*

PULVEROBOLETUS RAVENELII *(Berkeley and Curtis) Murrill, Mycologia 1:9. 1909*

Boletus ravenelii Berkeley and Curtis, Ann. Mag. Nat. Hist. II, 12:429. 1853.

Illustrations: See Microfiche No. 36

Smith, A. H., and H. D. Thiers, The Boletes of Michigan, pls. 149–150

Snell, W. H., and E. A. Dick, The Boleti of Northeastern North America, pl. 29.

Pileus 5–13 cm broad, convex to subglobose when young, becoming convex to broadly convex to plano-convex at maturity; surface viscid to subviscid but typically overlain with a dry veil, tomentose to pulverulent due to velar layer, glabrous to appressed fibrillose when veil fragments disappear, sometimes appearing fibrillose-scaly or eventually rimose-areolate when very old; background color typically pinkish to orange brown ("pale pinkish buff" to "pinkish cinnamon") with fibrils or streaks of reddish brown ("Hays russet" or "russet"), but background may be as pale as buff color ("light buff") or brown ("cinnamon"), veil elements colored intensely yellow ("wax yellow" to "strontian yellow") in young basidiocarps, unchanging or becoming a paler yellow ("light greenish yellow" to "greenish yellow") when old; margin incurved, entire but often appendiculate with veil fragments. *Context* 1–2 cm thick, whitish to pale yellow ("ivory yellow" to "pale pinkish buff"), slowly changing to blue when exposed, sometimes becoming brown or yellow when exposed. Taste somewhat acid; odor not distinctive.

Tubes 0.5–1.5 cm long, adnate, becoming slightly depressed to free with age, bright yellow ("primuline yellow" to "lemon yellow") when young, becoming darker yellow ("buffy citrine" to "olive lake" to "chamois") with age, turning blue to greenish blue when bruised, sometimes eventually changing to brown ("cinnamon-brown") to blackish; *pores* 0.5–1 mm broad, round to angular, concolorous with tubes, bluing when bruised.

Stipe 6–11 cm long, 0.7–1.5 cm thick at the apex, equal or occasionally tapering toward the apex or irregular in shape, solid, with white to pale yellow rhizoids at the base; surface dry to moist to occasionally subviscid, yellow ("primuline yellow"), typically covered with veil remnants colored bright yellow ("greenish yellow" to "strontian yellow" to "citron yellow"), sometimes appearing somewhat annulate. *Context* pale ochraceous to whitish except intensely yellow in base, unchanging or developing a pinkish tinge when exposed.

Spore print dark olive brown. *Spores* 8–11.2 X 4–6.4 μm, fusoid to ovoid to subellipsoid, ochraceous in KOH with a hyaline outer sheath, ochraceous tawny in Melzer's, smooth, moderately thick walled.

Basidia 24–35 X 10–14 μm, four-spored, hyaline, clavate. Hymenium typically yellow in KOH. *Hymenial cystidia* 30–56 X 5–12 μm, scattered to numerous, often somewhat embedded in the hymenium, hyaline to lemon yellow in KOH, clavate to fusoid to subcylindric, thin-walled.

Tube trama strongly divergent from distinct central strand, hyaline, gelatinous in KOH. *Pileus trama* interwoven, homogeneous. *Pileus cuticle* differentiated as a trichodermium of interwoven hyphae. Veil tissue composed of a mass of narrow, loosely interwoven hyphae staining yellow in KOH, no sphaerocysts apparent. *Stipe cuticle* similar to that of the pileus. *Clamp connections* absent.

Chemical reactions HNO_3—context pink, cuticle pink.

Habit, habitat, and distribution Solitary in humus in mixed forests of coastal regions. Most commonly found under oaks. Fruiting during fall and winter rainy season. As far as can be determined, this is the first report of this species from the state and possibly from the entire Pacific coastal area.

Material studied Del Norte County: Thiers 21540. Mendocino County: Ripley 899; Thiers 9331, 9350, 9439, 10636, 11023, 21398, 21507, 21528, 26896.

Observations It is very difficult to confuse this bolete with any other. It is readily recognized by the cottony, bright yellow veil, which, when young, envelops the entire basidiocarp. When older or after heavy rainstorms the veil, to a large extent, is likely to disappear; however, there are usually surviving fragments on the stipe or margin of the pileus that indicate its presence. The structure of the veil is interesting in that it does not possess any of the large, rounded cells (sphaerocysts) frequently seen in veils of other species.

Edibility unknown.

SUILLUS S. F. GRAY

Emended by Smith and Thiers
Contr. Toward a Monogr. of N. A. Sp. of Suillus, 1964

Suillus S. F. Gray, Nat. Arr. Brit. Pls. 1:646. 1821

Pileus convex to plane to occasionally umbonate or depressed; surface dry to moist to subviscid or viscid or glutinous, in California most frequently viscid, glabrous or streaked to appressed fibrillose to fibrillose-scaly, pale tan or buff to pallid vinaceous to yellow, unchanging with age or becoming yellow to brown; margin sometimes with conspicuous cottony roll (false veil), especially when young; context white or yellow, unchanging or becoming blue or very rarely reddening when exposed; taste and odor mild or rarely acid. Tubes 0.5–2 cm long, pallid to yellow during all stages of development; pores 0.5–3(5) mm broad, angular, concolorous, sometimes radiately arranged. Stipe central, equal to rarely clavate or ventricose, pallid to yellow; surface dry to subviscid or viscid, often noticeably glandulose; annulus present or absent. Spores olive brown in mass, subellipsoid to subcylindric to subfusoid, pale ochraceous in KOH, walls smooth, relatively thin. Cystidia solitary or clustered or fascicled, hyaline, becoming dark brown to almost black to occasionally vinaceous in KOH, clavate to cylindric. Hymenophoral trama divergent. Cuticle a trichodermium or ixotrichodermium. Clamp connections absent in California species. Forms mycorrhizal associations with conifers, especially pines and Douglas fir.

Suillus S. F. Gray

Key to Sections

1. Pileus dry to moist, fibrillose to fibrillose-scaly, or, if viscid, then stipe annulate and without glandular dots on the surface, at least when young	Sect. *Boletinus*
1. Pileus subviscid to viscid to glutinous; stipe without annulus or, if annulate, then surface with glandular dots	Sect. *Suillus*

SECTION BOLETINUS

Key to Species

1. Surface of pileus dry to moist, fibrillose to fibrillose-scaly (check young basidiocarps)	2
1. Surface of pileus viscid to glutinous; glabrous to streaked or appressed fibrillose	3
2. Pileus dull red to reddish brown to orange buff; surface fibrillose-scaly; viscid layer below scales well developed	*S. lakei* var. *lakei*
2. Pileus brick red to vinaceous tawny; scales prominent and more or less erect; viscid layer below scales poorly developed	*S. lakei* var. *pseudopictus*
3. Context of pileus changing to blue then fuscous when exposed	*S. lithocarpi-sequoiae*
3. Context of pileus not changing to blue, but may change to some other color and context of stipe may change to blue	4
4. Context of stipe, at least in the base, changing to blue when exposed	5
4. Context of stipe not changing to blue when exposed	*S. pseudobrevipes*
5. Annulus viscid to glutinous, outer surface bright orange in young basidiocarps; pileus brown to reddish brown, often staining greenish with age	*S. ponderosus*
5. Annulus moist to subviscid, pallid to white	6
6. Spores 7–9 μm long; pileus glabrous, cinnamon to orange cinnamon	*S. imitatus*
6. Spores 8–11 μm long; pileus pallid, buff to pale vinaceous, streaked or with closely appressed fibrils	*S. caerulescens*

SUILLUS CAERULESCENS *Smith and Thiers, Contr. Toward a Monogr. of N. Am. Sp. of Suillus p. 36. 1964*

Illustrations: See Microfiche No. 40

Smith, A. H., and H. D. Thiers, A Contribution Toward a Monograph of North American Species of Suillus, pl. 10.

Pileus 6–14 cm broad, broadly and shallowly convex when young, becoming plano-convex to plane with age; surface viscid, usually with patches of agglutinated fibrillose tomentum appearing as scattered squamules or streaks, frequently glabrous with age; color uniformly dull reddish to reddish brown ("vinaceous" to "ochraceous-tawny") on the disc and yellowish ("ochraceous-buff") toward the margin, squamules near the margin colored brown ("cinnamon"); margin at times appendiculate with partial-veil remnants. *Context* 10–20 mm thick, pale yellow ("ivory yellow"), unchanging or becoming dingy pink ("avellaneous"). Taste and odor mild.

Tubes 6–10 mm long, adnate to short decurrent, when young yellow ("primrose yellow" to "sulfur yellow" to "deep colonial buff"), unchanging or darkening slightly with age, staining dingy vinaceous brown ("avellaneous") when bruised; *pores* 1.5–2 mm long, ±1 mm wide, yellow ("honey yellow"), very irregularly angular, sometimes radially aligned but not always so, occasionally appearing compound.

Stipe 2.5–8 cm long, 2–3 cm thick at the apex, tapered toward the base or apex, or equal, solid; surface reticulate at the apex, glabrous to punctate above the annulus, glabrous to appressed fibrillose to matted fibrillose below, not glandular dotted, staining brown when handled, concolorous with the tubes above the annulus, becoming duller and often dingy brown below; annulus fibrillose, pallid to white but soon discoloring as the pileus, not strongly gelatinous. *Context* yellow (like the pileus margin), slowly staining blue in the base when exposed.

Spore print dingy cinnamon. *Spores* 8–11 X 3–5 μm, smooth, hyaline to ochraceous in KOH, pale orange buff to cinnamon buff in Melzer's, ellipsoid in face view, obscurely inequilateral in profile, thin-walled.

Basidia 20–30 X 5–8 μm, four-spored, clavate, yellowish in KOH and Melzer's. *Hymenial cystidia* of two types, the first 35–70 X 5–8 μm, cylindric to clavate with brown pigment around base of cluster, with content brown to hyaline in KOH, the other 22–28 X 5–8 μm, fusoid-ventricose and isolated to scattered, with refractive granules.

Tube trama hyaline, gelatinous, somewhat divergent near the subhymenium, with an obscure to distinct central strand, numerous large contorted laticifers present staining dark brown in KOH. *Pileus trama* interwoven, homogeneous, occasional laticifers present. *Pileus cuticle* up to 400 μm thick, differentiated as an ixotrichodermium with hyphae 8–15 μm wide, superimposed on a basal layer with mostly unbranched hyphae 3–7 μm wide, hyaline to pale ochraceous in KOH. *Caulocystidia* similar to hymenial cystidia. *Clamp connections* not seen.

Chemical reactions HNO_3–context and tubes yellow; H_2SO_4–context yellow, tubes reddish; $FeSO_4$–context and tubes gray; guaiac–cuticle, context and tubes blue.

Habit, habitat, and distribution Solitary to gregarious in soil under Douglas fir. Often abundant in the northern coastal forests, but not yet known from the higher elevations or from the more southerly portions of the state. Indications are that, at least in California, Douglas fir is the exclusive mycorrhizal host. Characteristically appearing early in the fall, *S. caerulescens* persists throughout the fall and winter season.

Material studied Del Norte County: Thiers 17645, 17731. Humboldt County: Thiers 13763, 14092, 14180, 17746, 24134, 24353. Marin County: Madden 625, 695; Thiers 11976, 24643. Mendocino County: Motta 58; Sundberg 68; Thiers 8439, 8751, 9042, 9511, 9630, 10532, 10703, 11054, 14115, 14609, 21519, 23062, 24204. Napa County: Sundberg 84; Thiers 10168, 10803, 11892. Nevada County: Thiers 26906. San Mateo County: Thiers 11820, 11916, 17909, 18176. Santa Cruz County: Largent 363, 389; Thiers 10782, 10793, 12035, 12299, 17937, 18533. Shasta County: Thiers 18207, 21586, 21673. Sonoma County: Thiers 9415, 10235, 12501, 18034. Trinity County: Thiers 10852.

Observations The large size of the cap, the glutinous and streaked pileus surface, which is never very darkly colored, the boletinoid arrangement of the tubes, the white to pallid annulus that is not glutinous, and the blue discoloration developing in the context of the basal part of the stipe are characters that distinguish this bolete. It is most likely to be confused with and certainly appears closely related to *S. ponderosus*, which, however, has an annulus that is not only glutinous but also colored bright yellow to reddish brown or orange, and a pileus that is typically glabrous and more deeply colored.

Edible, but of inferior quality.

SUILLUS IMITATUS *Smith and Thiers, Contr. Toward a Monogr. of N. Am. Sp. of Suillus, p. 40. 1964*

Illustrations

Smith, A. H., and H. D. Thiers, A Contribution Toward a Monograph of North American Species of Suillus, pl. 12.

Pileus 4–12 cm broad, obtuse to convex, expanding to nearly plane or the disc depressed, surface glutinous and glabrous; color brown to reddish brown ("orange cinnamon" to "cinnamon"). *Context* up to 2 cm thick, soft, pale yellow, slowly becoming dingy with a faint tawny cast where exposed (no definite color change). Taste mild; odor acidulous.

Tubes 7–15 mm deep, adnate or slightly depressed at the stipe, with decurrent lines up to 5 mm down the stipe apex, readily separable from pileus, yellow ("deep colonial buff") and becoming a little darker than honey yellow when older; *pores* 2–3 mm long, 1.5–2 mm wide, staining dull reddish brown, angular or irregular and elongated radially, the arrangement not obviously boletinoid.

Stipe 3.5–6 cm long, 1.5–2.5 cm thick, tapering abruptly at base, solid; surface at apex with decurrent tube lines, or faintly reticulate and smooth from there to the annulus, matted fibrillose and more or less ridged below the annulus, not glandular dotted anywhere; color yellow ("colonial buff") with tawny mottling above, pallid mottled cinnamon to pinkish cinnamon below, some finally discolored grayish brown to umbrinous; annulus band-like, well developed to faint, consisting of felty-tomentose soft material with some gluten adhering along the edges, colored white to dingy pallid and discoloring brownish. *Context* rather bright yellow, quickly becoming blue in basal half or third, elsewhere fading and becoming discolored like the pileus context.

Spore print color unknown. *Spores* 7–9 X 4–4.5 µm, narrowly elliptic in face view, subelliptic in profile, pale ochraceous in KOH, smooth, walls slightly thickened.

Basidia 18–23 X 5–6 µm, four-spored, hyaline to yellowish in KOH, clavate. *Hymenial cystidia* 36–54 X 7–10 µm, solitary or in bundles, subcylindric, narrowly clavate or submucronate, thin-walled, hyaline or with ochraceous to yellow-brown content, typically with amorphous bister (in KOH) pigment around the base of the bundle and at times over individual cystidia.

Tube trama of divergent hyphae, no laticifers seen. *Pileus trama* of interwoven hyphae and poorly differentiated laticiferous hyphae with considerable amorphous pigment particles scattered through the trama when revived in

KOH. *Pileus cuticle* an ixotrichodermium composed of repent narrow hyphae. *Caulocystidia* clavate to subcylindric, in fascicles, content ochraceous to tawny but lacking incrustations around the base or anywhere else. *Clamp connections* absent.

Chemical reactions unknown.

Habit, habitat, and distribution This species has not yet been recorded from California, but is known from both Oregon and Washington. In those states it occurs in spruce–fir forests and, since both of these hosts occur in the northern portion of California, it has been included with the expectation that it will eventually be discovered within the state.

Observations The description of this species is adapted from an earlier publication of Smith and Thiers. This fungus appears closely related to two rather common species in California. It has been confused, for example, with *Suillus lakei*, but *S. imitatus* has a glabrous pileus. In addition *S. lakei*, so far as is known, is associated only with Douglas fir. *Suillus caerulescens*, apparently the closest relative of *S. imitatus*, is distinguished by its longer spores, the presence of scattered, fibrillose scales on the pileus, and paler colors. *Suillus caerulescens* is also associated only with Douglas fir, at least in California.

Edibility unknown.

SUILLUS LAKEI *var.* **LAKEI** *(Murrill) Smith and Thiers, Contr. Toward a Monogr. of N. Am. Sp. of Suillus, p. 34. 1964*

Boletus lakei Murrill, Mycologia 4:97. 1912.

Ixocomus lakei (Murrill) Singer, Rev. Mycol. 5:6. 1940.

Boletinus lakei (Murrill) Singer, Farlowia 2:257. 1945.

Illustrations: See Microfiche No. 43

Smith, A. H., Mushrooms in Their Natural Habitats, Reel V.

Smith, A. H., Mushroom Hunter's Field Guide, pl. 46, p. 79.

Smith, A. H., and H. D. Thiers, A contribution Toward a Monograph of North American Species of Suillus, pls. 8, 9.

Pileus 7–14 cm broad when mature, convex to plano-convex when young, unchanging or becoming plane to shallowly depressed to highly irregular in outline at maturity; surface dry but with a gelatinous layer beneath the fibrils, scaly to squamulose during all stages of development, tips of scales frequently becoming free and somewhat recurved, rarely appearing glabrous when very old; color when young reddish to pink ("rufous" to "ferruginous" to "testaceous" to "vinaceous-tawny"), occasionally as pale as pale pink ("vinaceous-pink"), when older background color yellow ("straw yellow" to "barium yellow"), fibrils and scales reddish ("tawny" to "ochraceous-tawny" to "rufous"), occasionally with age becoming pallid ("pale olive buff" to "warm buff"); margin incurved when young, becoming decurved to highly eroded with age, frequently appearing appendiculate from partial-veil fragments. *Context* 1–2 cm thick, yellow ("barium yellow" to "mustard yellow"), changing to pinkish ("pale vinaceous" to "avellaneous" to "testaceous") when exposed. Taste and odor not distinctive.

Tubes 1–1.5 cm long, adnate to subdecurrent when young, typically decurrent with age, when young yellow ("amber yellow" to "pinard yellow"), when older dark yellow ("yellow ocher" to "old gold"), staining reddish ("testaceous") when bruised; *pores* 1–3 mm long, 1–2 mm broad, radially elongate, compound, concolorous with the tubes, angular.

Stipe 3–6 cm long, 1.5–2 cm thick at the apex, white mycelium at the base, solid; surface dry, sometimes obscurely reticulate at the apex, appressed fibrillose to obscurely fibrillose-scaly toward the base, concolorous with the tubes at the apex, ± concolorous with the pileus toward the base, broad, white annular zone in the apical region. *Context* concolorous with that of the pileus, changing to blue when exposed in old basidiocarps.

Spore print brown. *Spores* 8–11 X 3–4 μm, subellipsoid to subcylindric in face view, ventricose in side view, hyaline to pale ochraceous in KOH and Melzer's, smooth, thin-walled.

Basidia 25–35 X 9–12 μm, clavate, hyaline, two- and four-spored. *Hymenial cystidia* 45–60 X 6–12 μm, numerous to abundant, solitary on side of tubes but clustered on pores, incrusted, dark brown in KOH, cylindric to subclavate.

Tube trama divergent from a poorly defined mediostratum, hyaline, gelatinous in KOH. *Pileus trama* interwoven, homogeneous except for scattered laticiferous hyphae. *Pileus cuticle* differentiated as a basal gelatinous layer of interwoven hyphae subtending clusters of more or less erect, nongelatinous hyphae representing the scales. *Stipe cuticle* in apical portion composed of a layer of fertile basidia with scattered caulocystidia. *Clamp connections* absent or rare.

Chemical reactions KOH–cuticle olive; $FeSO_4$–context gray, tubes gray.

Habit, habitat, and distribution Scattered to gregarious in humus under Douglas fir. This species is more abundant in the coastal forests but collections have also been more abundant in the coastal forests but collections have also been made at higher elevations. Like several other species in the section *Boletinus* in California, it is apparently exclusively associated with Douglas fir.

Material studied Butte County: Ripley 1641. Del Norte County: Thiers 14050, 17643, 17734. Humboldt County: Thiers 14089. Marin County: Madden 745; Thiers 11806, 11975. Mendocino County: Thiers 8427, 8747, 8791, 8868, 9245, 9252, 9287, 9469, 9614, 9745, 10701, 11860, 14611, 21520, 23061. Napa County: Sundberg 184; Thiers 10175, 10800, 11887, 12482. San Mateo County: Thiers 11197. Santa Cruz County: Thiers 9706, 10958, 17934. Shasta County: Thiers 18206, 21678, 21700, 23088. Siskiyou County: Thiers 21692. Trinity County: Thiers 9405, 10843.

Observations This very distinctive species is easily recognized by the brick-red color of the pileus, the dry cuticle, and the conspicuous fibrillose scaly surface. This is one of the few *Suilli* with a typically dry pileus; however, care should be taken to examine young specimens, since in very old pilei the fibrillose scales will disappear and the inner gelatinous or viscid layer becomes very pronounced. Other species with a fibrillose cuticle are either very viscid or else belong to the section *Suillus*.

As has been indicated by Smith and Thiers, this is apparently the species that was earlier identified as *Suillus amabilis* by Slipp and Snell and others. According to the type description, however, the pileus of *S. amabilis* is glabrous and the context is pallid. Since the cuticle is so distinctive in *S. lakei*, it is difficult to accept the concept that the two species are synonymous. Singer, on the other hand, strongly believes the two are the same, and in an effort to further this belief recently visited what is possibly the type locality of *S. amabilis* in southern Colorado. While there he made a collection of *S. lakei* which, in his opinion, was ample proof that the two are identical. The logic of arriving at such a decision on this type of evidence is quite perplexing to this author, and it is hoped that it will not become common practice to reduce species to synonymy on the basis of geographical location.

Edible.

SUILLUS LAKEI *var.* PSEUDOPICTUS *Smith and Thiers, Contr. Toward a Monogr. of N. Am. Sp. of Suillus, p. 33. 1964*

Illustrations

Smith, A. H., and H. D. Thiers, A Contribution Toward a Monograph of North American Species of Suillus, pl. 7.

Pileus 5–11 cm broad, convex expanding to plano-convex to plane, to plane with an irregular uplifted margin at maturity; when young covered by conspicuous fibrillose scales, the fibrils or scales appressed near the margin, becoming somewhat squarrose on the disc, in age the yellow context showing between the scales; color of scales brick red to reddish brown ("brick red" to "vinaceous-tawny"), in age becoming grayish; margin typically appendiculate with fragments of the veil. *Context* 1–2 cm thick on the disc, yellow ("barium yellow" to "mustard yellow"), changing to pinkish vinaceous ("pale vinaceous" to "avellaneous" to "testaceous") when exposed. Taste and odor mild.

Tubes 1–1.5 cm long, adnate to slightly decurrent, bright yellow ("amber yellow" to "pinard yellow"), changing to dark yellow ("yellow ocher" to "old gold") with age; *pores* 1–3 mm long, 1–2 mm wide, concolorous, staining pinkish brown or darker when bruised, angular, radially elongate.

Stipe 3–6 cm long, 1.5–2 cm thick at apex, equal or pinched at the base, solid; surface dry, sometimes obscurely reticulate at the apex, yellow ("amber yellow" to "pinard yellow") with reddish streaks toward the base, base staining bluish when injured, veil submembranous, dry, thin, usually leaving a thin, sometimes evanescent annulus or with fragments remaining attached to the margin of the pileus. *Context* yellow, sometimes turning blue when exposed.

Spore print olive to olive brown. *Spores* 7–9 × 3.5–4 μm, pale ochraceous in Melzer's, greenish yellow to ochraceous in KOH, smooth, thin-walled, ellipsoid to subfusiform in face view, ventricose in profile.

Basidia 24–30 × 6–8 μm, two- and four-spored, clavate, hyaline to pale ochraceous in KOH. *Hymenial cystidia* 36–96 × 6–11 μm, numerous to abundant, typically occurring in bundles with incrustations of rust-brown pigment surrounding the base, content of many staining pale brown in KOH, clavate to more or less cylindric with obtuse to ovate-pointed apices.

Tube trama hyaline, gelatinous in KOH, divergent when young, appearing somewhat interwoven with age. *Pileus trama* interwoven, homogeneous, except for noticeable laticiferous hyphae, which stain dark brown in Melzer's and KOH. *Pileus cuticle* differentiated as a trichodermium of scales composed of masses of more or less erect hyphae 5–9 μm wide, no gelatinous subcuticular layer. *Stipe* *cuticle* in apical region composed of a layer of fertile basidia with scattered clusters of inconspicuous caulocystidia similar to those on the tubes. *Clamp connections* absent.

Chemical reactions unknown.

Habit, habitat, and distribution Scattered to gregarious in soil under Douglas fir. Known only from Jackson State Forest in Mendocino County. This variety is less common in the state than the typical one, and, as indicated above, is known only from a relatively small region in the northern coastal forest. It is, however, known from other states in the Pacific Northwest.

Material studied Mendocino County: Largent 169; Thiers 8206, 8282, 8782, 14606.

Observations This species differs from the typical variety in having a darker red pileus with more elaborate fibrillose scales on the surface; also, the inner gelatinous layer of the cutis is not as well developed, although there is considerable variation in this character.

Edible.

SUILLUS LITHOCARPI-SEQUOIAE *Singer, Mycologia 51:589. 1959*

Pulveroboletus lithocarpi-sequoiae Singer, The Agaricales in Modern Taxonomy, p. 734. 1962.

Pileus 7–14 cm broad, glabrous, obtuse expanding to broadly convex, glutinous overall, but soon dry, more or less rugulose, variegated cinnamon, reddish brown, and pale alutaceous (rarely uniformly colored). *Context* firm but soon soft, whitish to yellowish, staining blue then fuscous, in age lemon yellow throughout and then not bluing so readily. Taste mild; odor agreeable, fruity.

Tubes up to 8 mm deep (rather shallow), primuline to golden yellow, becoming dingy olive ochre to dusky olive brown, arcuate to short decurrent; pores concolorous with sides of tubes, almost round to obscurely boletinoid, sublamellate near pileus margin for a short distance, becoming compound and finally about 1 mm wide.

Stipe 8–11 cm long, 22–33 mm thick, equal or attenuated near base (not rooting), solid; surface pallid to dingy pale yellowish to brownish, apical part ferruginous-squamulose to furfuraceous, slightly ornamented in the same way below the annulus, ornamentation less pronounced downward; veil with the outer layer glutinous (copiously so), inner layer whitish to yellowish and membranous; annulus becoming appressed to stipe and lacerate to denticulate on both margins, at times oblique, superior.

Spore print color unknown. *Spores* 7.5–9.8 X 3.2–3.8 μm, apparently oblong in face view and subfusoid in profile, pale melleous, smooth, walls thin.

Basidia 29–30 X 6–7 μm, elongate, clavate, four-spored. *Hymenial cystidia* 48–84 X 4–10.5 μm, solitary or in fascicles, incrusted, pigment vinaceous in KOH, cylindric to somewhat ventricose. Subhymenium of irregular small elements, with incrusting pigment vinaceous in KOH.

Tube trama bilateral, gelatinous, mediostratum of yellowish somewhat interwoven hyphae. *Pileus cuticle* a thick gelatinous layer of hyphae 1–2.2 μm wide, some yellowish incrustations present. *Clamp connections* absent.

Chemical reactions unknown.

Habit, habitat, and distribution On the ground in mixed woods of *Sequoia* and *Lithocarpus* during the winter rainy season.

Observations This is a species of *Suillus* that has not been collected by the author; this description is taken from the original by Singer. It was described by him from collections made in Muir Woods National Monument in Marin County in 1958, where it is apparently very rare since year around collecting in that area since 1959 has failed to reveal it. There is some confusion regarding its relationship with other *Suilli* and boletes. Singer recently transferred it to the genus *Pulveroboletus*. However, it is the opinion of this author that it should remain in *Suillus* because of the presence of clustered cystidia, small pores, viscid pileus, and the presence of an annulus. Since no collections have been seen, no comments can be made regarding its affinities with other species.

Edibility unknown.

SUILLUS PONDEROSUS *Smith and Thiers, Contr. Toward a Monogr. of N. Am. Sp. of Suillus, p. 38. 1964*

Illustrations: See Microfiche No. 45

Smith, A. H., and H. D. Thiers, A Contribution Toward a Monograph of North American Species of Suillus pl. 11.

Pileus 9–25 cm broad, convex when young, becoming plane to irregular with age; surface glabrous or with scattered veil remnants near the margin, viscid to glutinous, somewhat streaked beneath the gluten; color reddish to dark brownish lavender ("vinaceous-brown" to "testaceous") to sometimes brown ("cinnamon") or yellow toward the margin and streaked with cinnamon; margin appendiculate from veil remnants. *Context* 2–3 cm thick, firm, yellow, unchanging or becoming pale vinaceous in spots. Taste mild; odor mild or acidulous.

Tubes up to 1.5 cm long, decurrent, dull yellow, unchanging when exposed; *pores* large, 1–3 mm in longest dimension, subangular, yellow, staining brownish.

Stipe 9–14 cm long, 3–6 cm thick, tapered toward the base, often pinched, solid; surface reticulate above the annulus from decurrent tubes, sometimes obscurely reticulate below the annulus, bright yellow above annulus, sordid rust brown when handled; annulus orange to white, membranous, superior, often stained reddish cinnamon on underside by a gelatinous layer. *Context* pale yellow throughout, greenish in base when exposed, with a faint pink tinge toward the apex.

Spore print brown. *Spores* 8–10 X 3.8–5 μm, hyaline to olive yellow in KOH, pale ochraceous in Melzer's, in face view ellipsoid to subfusoid to oblong, in side view somewhat inequilateral, smooth, thin-walled.

Basidia 26–31 X 7–10 μm, four-spored, clavate, hyaline in KOH. *Hymenial cystidia* 29–72 X 5–8 μm, solitary or clustered with dark yellow-brown pigment in and around the base, cylindric to clavate, or crooked, at times subfusoid, hyaline or with dark-brown pigment as revived in KOH.

Tube trama hyaline, gelatinous, somewhat divergent, hyphae 6–8 μm wide. *Pileus trama* interwoven, homogeneous, scattered laticifers present. *Pileus cuticle* a gelatinous layer of interwoven, hyaline hyphae, 2–4 μm wide, pale ochraceous in KOH. *Clamp connections* not seen.

Chemical reactions KOH—cuticle black, tubes black, context vinaceous; NH_4OH—context pale lavender, tubes lavender; HNO_3—tubes pale orange; H_2SO_4—context yellow, tubes red; $FeSO_4$—context and tubes gray.

Habit, habitat, and distribution Scattered to cespitose in soil under Douglas fir in coastal forests of northern California.

Material studied Marin County: Madden 693, 780, 869; Thiers 10040, 18124. Mendocino County: Peters 174, 214; Thiers 8163, 8207, 8279, 8382, 8472, 8612, 8762, 8776, 8785, 8821, 8867, 8874, 9242, 9251, 9284, 9338, 9354, 10531, 10570, 14107, 14143, 14144, 14610, 17899, 21511, 23081. Napa County: Thiers 11897, 21808. Santa Cruz County: 17941. Sonoma County: Motta 94.

Observations This is the largest species of *Suillus* known from California, with some pilei measuring up to 25 cm in diameter. Smaller-sized basidiocarps, however, are much more commonly found. *Suillus ponderosus* is further characterized by the bright-reddish-brown color of the pileus, the glabrous surface of the pileus, and the viscid to glutinous annulus, which is bright orange to orange yellow in young basidiocarps. This species appears most closely related to *S. caerulescens,* and they both form mycorrhizal associations with Douglas fir. Older basidiocarps of the two species are similar in appearance; however, younger developmental stages can be separated by using the characters enumerated above.

Edible.

SUILLUS PSEUDOBREVIPES *Smith and Thiers, Contr. Toward a Monogr. of N. Am. Sp. of Suillus, p. 103. 1964*

Boletus granulatus var. *albidipes* Peck, Ann. Rept. N.Y. State Mus. 54:168.1901.

Illustrations: See Microfiche No. 46

Smith, A. H., and H. D. Thiers, A Contribution Toward a Monograph of North American Species of Suillus, pl. 16.

Pileus 8–15 cm broad when mature, convex to broadly convex when young, becoming plano-convex to plane at maturity; surface viscid to glutinous, glabrous, occasionally with patches of white veil fragments near the margin, often appearing streaked under the gluten; color dark yellow brown ("buckthorn brown" to "clay color"), becoming darker brown ("cinnamon-brown" to "Prout's brown") with age; margin incurved, becoming broadly decurved with age, often appendiculate from partial-veil fragments. *Context* 1–2 cm thick, white to pale yellow, unchanging when exposed. Taste and odor not distinctive.

Tubes 0.5–1 cm long, adnexed to arcuate-decurrent to decurrent, yellow ("naphthalene yellow" to "baryta yellow") when young, becoming dark yellow ("old gold" to "colonial buff") when older, unchanging when exposed; *pores* small, ±0.5 mm, angular, concolorous with tubes, unchanging when bruised.

Stipe 3–6 cm long, 1–2 cm thick, equal or tapering toward the base, solid; surface dry, white to pallid when young, typically becoming yellow with age, when young appearing glabrous with only obscure glandulae, with age glandulae often become brown and much more conspicuous; annulus distinct, median, fibrillose, sometimes sheathing lower portion of stipe, pallid or pale brownish lavender ("avellaneous"). *Context* white to whitish, unchanging when exposed.

Spore print pale brown. *Spores* 7–9 X 2.5–3.5 µm, hyaline to pale ochraceous in KOH and Melzer's, ellipsoid to subcylindric in face view, slightly inequilateral in side view, smooth, thin-walled.

Basidia 15–21 X 6–9 µm, clavate, four-spored, hyaline in KOH. *Hymenial cystidia* 21–35 X 7–10 µm, scattered, more abundant near the pores, clustered, hyaline, staining dark brown in KOH, clavate to subcylindric.

Tube trama gelatinous in KOH, hyaline, divergent. *Pileus trama* homogeneous, interwoven. *Pileus cuticle* differentiated as a tangled ixotrichodermium of narrow hyphae, 3–6 µm wide. *Stipe cuticle* interwoven, caulocystidia similar to those in the hymenium. *Clamp connections* absent.

Chemical reactions unknown.

Habit, habitat, and distribution Scattered to solitary in humus, usually under beach or lodgepole pines. Although this fungus is apparently very rare within the state, collections have been made in both the coastal and Sierra Mountain ranges.

Material studied Madera County: Thiers 20899. San Bernardino County: Thiers 24875. Santa Cruz County: Thiers 21703. Tuolumne County: Thiers 21304.

Observations This very interesting species appears rather closely related to several different groups of *Suilli.* The poorly developed, inconspicuous glandulae on the stipe suggest an affinity with *S. brevipes;* however, the presence of the annulus makes it distinct. *Suillus albidipes* and *S. glandulosipes* also have a partial veil, but the veil remnants adhere to the pileus margin rather than to the stipe as in *S. pseudobrevipes.* Furthermore, the pileus of *S. pseudobrevipes* is usually quite dark reddish brown in color, and no pallid or white stage of the basidiocarps is known. *Suillus luteus* might also be confused with this species, but can be separated by its distinctly glandulose stipe and usually well developed annulus, with a lavender or purple layer or band on the outer surface.

Edibility undetermined.

Suíllus S. F. Gray

SECTION SUILLUS

Key to Species

1. Context changing to blue when exposed		2
1. Context unchanging or not becoming blue when exposed		3
2. Surface of stipe distinctly reticulate	*S. reticulatus*	
2. Surface of stipe not reticulate	*S. tomentosus*	
3. Veil or false veil present, either forming an annulus or existing as a noticeable role of tissue on the pileus margin (check young basidiocarps)		4
3. Veil or false veil absent; margin glabrous or with scattered fibrils		18
4. Annulus typically present		5
4. Annulus typically absent		10
5. Pores very large, up to 5 mm broad; compound and often lamellate; annulus evanescent; pileus yellow to dark brown		6
5. Pores not more than 2 mm broad, usually smaller; annulus more or less persistent; pileus tan to pale brown to olive brown to dark brown		7
6. Pileus bright yellow; stipe short, often somewhat eccentric	*S. megaporinus*	
6. Pileus yellow brown; stipe well developed, not eccentric	*S. riparius*	
7. Stipe with noticeable glandular dots on the surface		8
7. Stipe without noticeable glandular dots	*S. pseudobrevipes* (See Boletinus Sec.)	
8. Pores 1 mm or more broad; cap more or less umbonate; pileus pale brown to tan or pale olive brown	*S. umbonatus*	

Couplet	Species	Go to
8. Pores less than 1 mm broad; pileus usually some shade of brown		9
9. Annulus with a vinaceous gray to purplish layer or zone on the outer surface; spores 7–9 μm long	*S. luteus*	
9. Annulus lacking a vinaceous or purplish zone; spores 9–11 μm long	*S. subolivaceus*	
10. Pores large, up to 5 mm broad, often lamellate; annulus evanescent		11
10. Basidiocarp not with above characters		12
11. Pileus bright yellow; stipe short, often eccentric	*S. megaporinus*	
11. Pileus yellow brown; stipe well developed, not eccentric	*S. riparius*	
12. Pileus white then gray, becoming ochraceous brown when mature; associated with Monterey and knobcone pines	*S. pungens*	
12. Basidiocarps and/or mycorrhizal host not as above		13
13. Pileus white, soon becoming chocolate brown; associated with sugar pine	*S. brunnescens*	
13. Basidiocarps and/or mycorrhizal host not as above		14
14. Pileus dark cinnamon brown; stipe white when young, becoming ochraceous when mature	*S. borealis*	
14. Basidiocarp not as above		15
15. Pores 1–2 mm broad; pileus dingy yellow; associated with white pine (*Pinus monticola*)	*S. sibiricus*	
15. Basidiocarps and/or mycorrhizal host not as above		16
16. Glands prominent on stipe during all stages; associated with Bishop pine	*S. glandulosipes*	
16. Basidiocarps and/or mycorrhizal host not as above		17
17. Pileus yellow to rust color when young, often spotted or mottled; spores 8–10 × 3–4 μm	*S. volcanalis*	

17. Pileus white to pallid to pale vinaceous when young; spores 6–8 × 2.5–3 μm	*S. albidipes*	
18. Pileus surface conspicuously fibrillose-scaly		19
18. Pileus surface glabrous, may appear spotted or streaked from gluten with age		21
19. Pileus brown; pores large; associated with ponderosa pine	*S. californicus*	
19. Basidiocarps and/or mycorrhizal hosts not as above		20
20. Taste mild	*S. fuscotomentosus*	
20. Taste harsh, unpleasant	*S. acerbus*	
21. Stipe without conspicuous glandular dots		22
21. Stipe with noticeable glandular dots		23
22. Stipe clavate to ventricose; yellow to olive to olive buff to pale brown	*S. monticolus*	
22. Stipe more or less equal, often short, white to pallid	*S. brevipes*	
23. Context white, pores radiately arranged	*S. punctatipes*	
23. Context yellow, pores not radiately arranged	*S. granulatus*	

SUILLUS ACERBUS *Smith and Thiers, Contr. Toward a Monogr. of N. Am. Sp. of Suillus, p. 103. 1964*

Illustrations: See Microfiche No. 37

Pileus 8–13.6 cm broad when mature, when young convex, becoming plane to plano-convex with a shallow to moderately deep depression on the disc, often undulating to highly irregular toward the margin; surface viscid, glabrous to fibrillose or fibrillose-scaly, often streaked with gluten when older; color highly variable, reddish to brown ("tawny" to "ochraceous-tawny" to "hazel" to "buckthorn brown" to "cinnamon" to "cinnamon-buff"), in some these colors appear overlain with grayish pigments, giving a dark brown or gray appearance, with age unchanging or fading to light brown ("cinnamon-buff" to "buffy brown"), or often a slightly paler mixture of some of the above colors, frequently with a greater concentration of darker pigments toward the margin; margin entire, incurved, glabrous. *Context* 1.5–2.5 cm thick on the disc, white, staining yellow in irregular areas and pinkish brown ("avellaneous" to "testaceous") around larval tunnels. Taste unpleasant, somewhat harsh and acid; odor not distinctive.

Tubes 0.5–1 cm long, adnate to arcuate-decurrent, typically becoming decurrent with age, when young white to pale yellow ("colonial buff" to "deep colonial buff") to less frequently pallid ("pale pinkish buff"), becoming yellow ("reed yellow" to "primrose yellow" to "chamois") with age, not changing when bruised or exposed; *pores* not exceeding 1 mm broad, angular, concolorous, unchanging when bruised.

Stipe 4–10 cm long, 1.5–2.5 cm thick at the apex, equal to slightly enlarged at the base, straight or frequently bent or crooked, solid; surface dry, glabrous but covered with large, inconspicuous to conspicuous glandulae that become brown ("cinnamon" to "ferruginous") with age or when handled, some obscure reticulation apparent at the apex and often with irregular longitudinal lines forming a crude network toward the base, no true raised reticulation present; color white to very pale yellow ("ivory yellow"), changing with age or when handled to yellow ("Naples yellow" to "amber yellow"), not changing in same intensity in all parts of the stipe; no annulus. *Context* white, typically staining yellowish when exposed.

Spore print dull brown. *Spores* 9.3–12 X 3.3–4 μm, hyaline to pale ochraceous in KOH, smooth, thin-walled, cylindric to subellipsoid in face view to subventricose in profile.

Basidia 24–29 X 5–8 μm, hyaline, clavate, contents granulose in KOH, four-spored. *Hymenial cystidia* 30–46 X 7–10 μm, fasciculate, scattered to very rare, found only near the pores of the tubes, basal portion often staining dark brown in KOH, apical portion remaining hyaline, occasionally staining vinaceous; no solitary cystidia seen.

Tube trama hyaline, divergent from a distinct mediostratum, subgelatinous, hyphae up to 7 μm wide. *Pileus trama* interwoven, homogeneous. *Pileus cuticle* differentiated as an interwoven broad (300 μm) ixotrichodermium, contents of hyphae appear brownish in KOH, outermost hyphae appearing somewhat differentiated and incrusted, hyphae up to 3.5 μm wide; hypodermium well differentiated and the hyphae incrusted, interwoven and dark brown in KOH. *Stipe cuticle* gelatinous with massive clusters of darkly staining (brown to dark vinaceous) caulocystidia similar to the other cystidia. *Clamp connections* absent.

Chemical reactions KOH–cuticle red, context reddish, tubes black; NH_4OH–context vinaceous, tubes red; HNO_3 and HCl–context pale yellow; H_2SO_4–tubes dark red; $FeSO_4$–context and tubes pale gray.

Habit, habitat, and distribution Gregarious to sometimes cespitose in soil under Monterey pines. Known only from the San Francisco Bay Area; sometimes abundant in the Presidio in San Francisco. It has not been found associated with any other of the numerous pines or conifers of the coastal forest.

Material studied Contra Costa County: Thiers 9687. Marin County: Shervanick 358. San Francisco County: Gleason 1-5-65; Hanks 11-10-64; Jordan 1019; Thiers 9685, 9686, 24518, 24537, 26887, 26962, 27065. San Mateo County: Thiers 11773, 12136; Torbett 11-16-64.

Observations This species is characterized by the brown to reddish-brown pileus, which appears granulose or fibrillose to tomentose, the absence of any type of veil, the unpleasant taste, and the relatively long spores. *Suillus pungens,* which also forms mycorrhizal associations with Monterey pine, shows a partial veil and white to gray color in the pileus when young. *Suillus acerbus* is possibly most clearly related to *S. granulatus,* which has smaller spores and a mild taste.

Edibility unknown.

SUILLUS ALBIDIPES *(Peck) Singer, Farlowia 2:45. 1945*

Boletus albidipes Peck, N.Y. State Mus. Bull. 57:22. 1912. (Not *Boletus granulatus* var. *albidipes* Peck, Ann. Rept. N.Y. State Mus. 54:168. 1901.)

Suillus albidipes (Peck) Snell, Mycologia 37:378. 1945.

Suillus granulatus ssp. *albidipes* (Peck) Snell and Dick, Mycologia 53:232. 1961.

Illustrations: See Microfiche No. 38

Smith, A. H., Mushroom Hunter's Field Guide, pl. 56a.

Smith, A. H., and H. D. Thiers, A Contribution Toward a Monograph of North American Species of Suillus, pl. 32.

Smith, A. H., and H. D. Thiers, The Boletes of Michigan, pls. 22, 23.

Snell, W. H., and E. A. Dick, The Boleti of Northeastern North America, pl. 20.

Pileus 5–12 cm broad when mature, convex becoming broadly convex to plano-convex to undulating and highly irregular in outline; surface glutinous to viscid; color white or pallid to pale vinaceous ("vinaceous-buff") when young, becoming vinaceous to brown ("vinaceous-cinnamon" to "cinnamon") and somewhat variegated or streaked when older; margin incurved, at first decorated by a dry cottony roll of whitish to vinaceous-buff material representing a false veil, which collapses and often disappears, but may form pallid patches along the margin in mature specimens. *Context* 1–2 cm thick, white, unchanging or slowly becoming yellow when exposed. Taste and odor not distinctive.

Tubes about 4–8 mm long, adnate, becoming adnexed to depressed, with drops of exudate, white to pale yellow ("ivory yellow") when very young, soon becoming yellow ("primuline yellow"); *pores* round, minute (±three per millimeter), yellow, typically not staining or becoming pale vinaceous when bruised.

Stipe 3–8 cm long, 1–2.5 cm thick, equal, bulbous, or tapered or pinched at the base, solid; surface white at first and not glandular dotted, in age slowly yellow above and reddish brown below with very fine granular dots toward the base; no annulus. *Context* white to yellowish within, with age cortex of midportion reddish, bright yellow ("lemon chrome") above, dingy brown in the base, unchanging when exposed.

Spore print dull brown. *Spores* 6.6–8.8 × 2.5–3 μm, thin-walled, greenish hyaline in KOH, yellowish in Melzer's, subellipsoid to ellipsoid in face view, oblong to obscurely inequilateral in side view, smooth, thin-walled.

Basidia 15–20 X 4–7 μm, clavate, four spored, hyaline. *Hymenial cystidia* 26–35 X 8–11 μm, clavate, brown in KOH, rare, clustered mostly near the pore, with dingy yellow brown incrusting pigment.

Tube trama divergent, hyaline in KOH, gelatinous in KOH. *Pileus trama* interwoven, homogeneous. *Pileus cuticle* differentiated as an ixotrichodermium, ochraceous in KOH and Melzer's, hyphae 3–6 μm wide. *Stipe cuticle* with caulocystidia and caulobasidia. *Clamp connections* absent.

Chemical reactions KOH—context pink, then lavender; $FeSO_4$—context gray.

Habit, habitat, and distribution Scattered to gregarious in humus under lodgepole pine. This species is known only from the Huntington Lake area in Fresno County and from the vicinity of Calaveras Big Trees State Park in Calaveras County.

Material studied Calaveras County: Thiers 21239. Fresno County: Thiers 13416, 13444.

Observations As has been indicated by Smith and Thiers, there is considerable confusion concerning this fungus due to the fact that Peck used this name for two different fungi. In our concept the name should be restricted to those basidiocarps lacking an annulus but possessing a distinct roll of veil tissue on the pileus margin. Obviously, this species is closely related to *S. granulatus*, but the presence of a partial veil as indicated by the cottony roll on the margin of the pileus makes it distinct. This character is sometimes absent in old basidiocarps.

Suillus albidipes is somewhat similar to *S. glandulosipes*, which, however, is only associated with Bishop and beach pine. Also, *S. glandulosipes* shows prominent glandulae on the stipe during all stages of development and is not white when young. *Suillus pungens*, in the San Francisco Bay Area, might be confused with *S. albidipes*, but *S. pungens* has a very distinctive color pattern in the pileus as the basidiocarp develops, a harsh, unpleasant taste, larger spores, and is typically associated with Monterey pine.

Edible.

SUILLUS BOREALIS *Smith, Thiers, and Miller, Lloydia 28:123. 1965*

Pileus 7–11 cm broad when mature, convex when young, becoming broadly convex to plano-convex to plane or shallowly depressed at maturity, often with undulating or highly irregular outlines when very old; surface very viscid to glutinous, glabrous, but some pilei appearing streaked from gluten; color brown ("orange-cinnamon" to "vinaceous-cinnamon" to "cinnamon"), sometimes dark brown ("Mikado brown" to "Verona brown") on the disc, fading to pale brown ("light pinkish cinnamon" to "pinkish buff") on the margin; margin strongly incurved, typically strongly appendiculate when young, but glabrous in some pilei when older, sterile. *Context* 1–2 cm thick, when young whitish, becoming yellow with age, changing to yellow when exposed in young basidiocarps. Taste and odor mild.

Tubes up to 1 cm long, shorter near the margin, adnate to broadly and shallowly depressed, yellow ("old gold"), unchanging when exposed; *pores* ±1 mm broad, angular, concolorous, unchanging when bruised.

Stipe 2–4 cm long, 1.5–2 cm thick at the apex, equal, solid to stuffed; surface dry, glandulae not strongly apparent when young but becoming so with maturity; color white when young, becoming reddish brown ("tawny" to "ochraceous-tawny"), frequently bright yellow at the apex, sometimes staining pale brown at the base; no annulus. *Context* white to yellow, unchanging when exposed.

Spore print brown. *Spores* 6–8.5 X 3–5 µm, hyaline in KOH, yellowish in Melzer's, subellipsoid to subcylindric, smooth, thin-walled.

Basidia 25–31 X 8–11 µm, clavate, four-spored, hyaline. *Hymenial cystidia* rare to absent on sides of tubes, abundant to crowded on the pores, dark brown in KOH and Melzer's, heavily incrusted, occasional hyaline clusters with only the base discoloring in KOH; individual cystidia 30–54 X 7–13 µm, cylindric to clavate, thin-walled.

Tube trama strongly divergent from a distinct mediostratum, subgelatinous in KOH, hyaline. *Pileus trama* interwoven, homogeneous, hyaline to pale brown in KOH. *Pileus cuticle* differentiated as a layer of interwoven to tangled hyphae 3–4 µm wide, gelatinous, hyaline to pale brown in KOH, rust brown in Melzer's. *Stipe cuticle* differentiated as a layer of gelatinous hyphae with numerous large fascicles of cystidia staining dark brown in KOH. *Clamp connections* absent.

Chemical reactions KOH—context vinaceous; $FeSO_4$—context gray.

Habit, habitat, and distribution Gregarious to subcespitose in soil in mixed pine–oak woods. Known only from a single collection made under pines in the vicinity of Willow Creek in northern California. This species was originally described from Idaho but is now known from throughout the Pacific northwest. It appears to be associated with five-needle pines, but further data are necessary before its mycorrhizal host can be accurately determined.

Material studied Humboldt County: Thiers 14266.

Observations This species is characterized by the presence of a veil but no annulus, the very dark brown color of the pileus, and the white to pallid stipe, at least when young, with obscure glandulae on the surface. There are no other species in California with which it is likely to be confused. In other areas it might be confused with *S. luteus,* but that species has a distinct annulus. *Suillus pseudogranulatus,* a species known only from the southeastern portion of the United States, also seems to be somewhat similar, but is much paler in color and also has an annulus or at least a fibrillose zone on the stipe.

Edibility unknown.

SUILLUS BREVIPES *(Peck) Kuntze, Rev. Gen. Pl. 3:535. 1898*
Boletus brevipes Peck, Ann. Rept. N.Y. State Mus. 38:110. 1885.

Illustrations: See Microfiche No. 39

Smith, A. H., The Mushroom Hunter's Field Guide, p. 89, pl. 55.

Smith, A. H., and H. D. Thiers, A Contribution Toward a Monograph of North American Species of Suillus, pl. 42.

Smith, A. H., and H. D. Thiers, The Boletes of Michigan, pl. 28.

Snell, W. H., and E. A. Dick, The Boleti of Northeastern North America, pl. 21.

Pileus 5–10 cm broad when mature, globose to convex becoming broadly convex to plano-convex to plane; surface glabrous, glutinous, in wet weather the gluten often 2 mm deep; color evenly dark vinaceous-brown ("Natal brown" to "army brown") when young, gradually becoming paler to dull cinnamon ("sayal brown" to "cinnamon"), sometimes dingy yellow ocher in age; margin incurved, becoming decurved to plane, entire, in buttons faintly white tomentose, but veil absent and no distinct roll of white cottony tissue present. *Context* 1–2 cm thick, white when young, becoming yellow in age, soft, unchanging when bruised. Odor and taste not distinctive.

Tubes 4–10 mm long, adnate to decurrent, yellow ("honey yellow"), darker and more olivaceous in age; *pores* 1–2 mm, round, not elongating radially, when young pale dingy yellow.

Stipe 2–6 cm long, 1–2 cm thick, equal, solid; surface white, becoming pale yellow, unpolished to pruinose under a lens and when young lacking glandular dots, grandulae sometimes visible in age but never well developed; no annulus. *Context* white, becoming yellow in cortex or apex.

Spore print brown. *Spores* 7–10 X 3–4 μm, narrowly elliptic to oblong, obscurely inequilateral in side view, smooth, pale yellowish in Melzer's and KOH.

Basidia 18–24 X 5–6 μm, four-spored, clavate, hyaline in KOH, yellowish in Melzer's. *Hymenial cystidia* clustered, with amorphous brown pigment surrounding the base of the cluster, individual cystidia 35–50 X 6–9 μm, cylindric to clavate, contents hyaline or either partly or entirely brown from coagulated pigment.

Tube trama divergent, gelatinous. *Pilcus trama* homogeneous, interwoven. *Pileus cuticle* differentiated as a thick ixotrichodermium compsed of narrow

(4–6 μm) hyphae. *Caulocystidia* like the hymenial cystidia, but the bundles not numerous. *Clamp connections* absent.

Chemical reactions According to Singer: KOH—dark gray on the cuticle, gray to dark brown on the tubes, lilac on the context; NH_4OH—purplish on the cuticle; H_2SO_4—rust ochraceous on the pores.

Habit, habitat, and distribution Solitary to scattered or gregarious in soil under two- and three-needle pines. This bolete occurs throughout the pine forests of the state, and is probably the most abundant and widely distributed *Suillus* within the state. It is typically associated with beach pine in the coastal forests and with lodgepole and occasionally ponderosa pine in the mountains.

Material studied Alpine County: Thiers 20805. El Dorado County: Thiers 20760, 23633. Fresno County: Thiers 13366. Humboldt County: Thiers 17728. Madera County: Thiers 20843, 23614. Marin County: Shervanick 380. Mariposa County: Thiers 20976. Mendocino County: Thiers 8298, 8777, 8862, 9302, 9334, 9353, 9474, 9814, 11073, 18390, 24236. Mono County: Thiers 20964. Nevada County: Thiers 13130, 13160, 13855, 26908. Placer County: Sundberg 336. Riverside County: Thiers 24999. Santa Cruz County: Thiers 10773, 10946. Shasta County: 21594, 21611, 21698, 21701, 23091. Sierra County: Sundberg 345; Thiers 13224, 21165. Siskiyou County: Thiers 21688. Tuolumne County: Thiers 13273.

Observations *Suillus brevipes* is recognized by the deep-reddish-brown pileus, which is glabrous and typically very viscid, the absence of a veil, and the usually short, pallid stipe, which has very obscure glandulae, at least when young. This is a highly variable fungus, and one should expect to find basidiocarps in which the stipe is considerably elongated and in which the color of the cap is quite variable. Particularly noticeable is the tendency of the pileus to fade and assume a noticeably streaked appearance when older or when exposed to prolonged cold or dry weather.

This fungus is probably most closely related to *S. granulatus* and *S. glandulosipes,* but these species typically have stipes in which the glandulae are strongly developed during all stages of development, and much paler or at least differently colored pilei.

Edible.

SUILLUS BRUNNESCENS *Smith and Thiers, Contr. Toward a Monogr. of N. Am. Sp. of Suillus, p. 91. 1964*

Illustrations

Smith, A. H., and H. D. Thiers, A Contribution Toward a Monograph of North American Species of Suillus, pls. 38, 39.

Pileus 4–15 cm broad, convex to nearly plane, surface glutinous to viscid; color at first dull white, the gluten soon changing to chocolate color, but often showing a lilac-brown tone in the darkening process, often appearing streaked from the gluten; herbarium specimens dark brown ("Natal brown") overall. *Context* thick, white when young, in age yellowish in apex of stipe and along the tubes or finally yellowish throughout, not staining when cut or bruised. Taste and odor mild.

Tubes up to 15 mm deep in large caps, mostly less than 10 mm deep, pale ochraceous, adnate to decurrent, not staining when bruised; *pores* small (about two per millimeter), pale ochraceous and not staining when bruised.

Stipe 3–6(8) cm long, 1–2.5(3) cm thick, solid; surface white at first and unpolished, apical region becoming yellowish with small glandular dots, base where handled staining vinaceous and then brownish like the pileus; veil thin, membranous, white staining vinaceous gray, breaking away from the stipe and collapsing on cap margin, in age all traces have disappeared. *Context* white within at first but slowly yellowish, especially in apex.

Spore print unknown. *Spores* 6.6–8.8 X 2.8–3.2 μm, oblong in face view, in profile oblong to obscurely inequilateral, smooth, thin-walled, hyaline to dingy yellow brown in KOH, pale yellow brown in Melzer's.

Basidia 18–25 X 4.4–6.6 μm, four-spored, clavate, hyaline to yellowish in KOH and often with amorphous granules, in Melzer's pale yellowish to brownish. *Hymenial cystidia* 40–70 X 8–12 μm, in fascicles with the bundles surrounded by incrusting pigment more or less bister in KOH, content hyaline to dingy yellow brown, subcylindric to fusoid-ventricose, thin-walled, smooth or with some incrusting debris.

Tube trama with a central strand of parallel, hyaline to dingy-yellow-brown hyphae (in KOH and in Melzer's), with hyaline, gelatinous hyphae diverging to the hymenium, the hyphae of the central strand (4)6–10(12) μm wide, walls thin, hyaline to yellowish in Melzer's. *Pileus cuticle* a gelatinous trichodermium of hyphae 4–9 μm wide, containing numerous dark-brown granules in KOH, the walls soon almost completely gelatinized. *Caulocystidia* similar to hymenial cystidia, in bundles and heavily incrusted around the base of the bundle with bister material. *Clamp connections* absent.

Chemical reactions KOH—context vinaceous lilac on the cuticle, slowly changing to greenish in places.

Habit, habitat, and distribution Scattered to gregarious under sugar pine. Known as far south as Grants Pass, Oregon.

Observations This species has never been reported from California. It is included, however, because it has been found nearby in the vicinity of Grants Pass in Oregon where it was associated with sugar pine. This description is adapted from the Smith and Thiers monograph of North American species of *Suillus.* As indicated by them, this species is distinguished by the white pileus when young, which changes to brown with age and stains brown when bruised, the tendency of the veil to remain attached to the pileus margin, and the obscure glandulae on the stipe. It appears to be closely related to *S. albidipes,* a species occasionally found within the state.

Edibility unknown.

SUILLUS CALIFORNICUS *(Murrill) Thiers comb. nov.*

Rostkovites californicus Murrill, Mycologia 7:44. 1915.

"Pileus thick, convex to plane, solitary, 6–9 cm broad; surface smooth; conspicuously subtomentose, brown, margin concolorous, entire, rather thick; context thick, fleshy, flavous, unchanging; taste mild; tubes adnate, plane or slightly convex in mass, 4–6 mm long, yellow, exuding drops which blacken with age, mouths large, angular; spores oblong-ellipsoid, smooth, yellowish brown, 7–8 X 3.5–5 μ [μm]; stipe subequal or bulbous, smooth, yellow, with black dots, unchanging, solid, yellow within, unchanging, 3–6 cm long, 1.5–2 cm thick.

"Type collected on the ground in pine woods in Grass Valley, California, Nov. 12, 1914, H. S. Yates and F. H. Bolster 251 (Herb. N. Y. Bot. Gard.). Excellent field notes accompany the specimens. The species is strikingly different from other members of the genus in having a conspicuously subtomentose surface resembling that of *Ceriomyces communis.*"

The above is a copy of the type description. No fresh basidiocarps of this species have been found. Through the courtesy of the New York Botanical Garden, the type has been examined. The following notes were included with the type collection.

"Number 251.

"*Rostkovites californicus* sp. nov."

(The first name given by the collector was *Rostkovites subaureus* but that species epithet had been struck through and *californicus* substituted.)

"Locality—Grass Valley, Calif.

"Date—November 12, 1914.

"Habitat—On ground in pine woods.

"Pileus—diameter 6–9 cm, subconvex to plane, subtomentose, shiny when fresh, color brown.

"Context fleshy, light yellow unchanging, taste mild.

"Tubes adnate; yellow; 4–6 mm long; mouths large, angular.

"Spores—7–8 X 3.5–4 μ [μm], brown.

"Stipe—Even or somewhat bulbous. 3–6 cm long by 1.5–2 cm in diameter. Color yellow with black granulations. Context of stipe yellow, unchanging. Base somewhat fibrous.

"Remarks—This species seems not to be affected by insects.

"Collector—F. H. Bolster.

"Sent to Murrill, Dec. 11, 1914.

"Notes by Yates."

A microscopic examination of the type collection (which consists of only one half a basidiocarp) yielded the following data on anatomical features.

Spores 9–12 X 3.5–5 μm, pale brown in KOH, some dextrinoid in Melzer's, ellipsoid to subovoid, inequilateral and ventricose in profile, smooth, thin-walled.

Basidia 24–29 X 7–9 μm, four-spored, clavate, hyaline in KOH, but hymenium pale ochraceous in KOH. *Hymenial cystidia* 35–46 X 7–10 μm, moderately thick walled, fascicled; dark brown in KOH, dextrinoid in Melzer's; more abundant at pores, some hyaline cystidia in clusters, clavate to somewhat irregular in outline, numerous solitary cystidia breaking free from the clusters and floating in the mount.

Tube trama strongly divergent, subgelatinous, hyaline in KOH, hyphae 4–6 μm wide. *Pileus trama* homogeneous, ochraceous in KOH, bright red in Melzer's. *Pileus cuticle* differentiated as a layer of narrow, interwoven hyphae that are subgelatinous, no pileocystidia or fascicles of erect hyphal tips seen, hyphae 3–5 μm wide, pale ochraceous in KOH, slightly darker in Melzer's. *Stipe cuticle* differentiated as a layer of fertile basidia with fascicles of caulocystidia interspersed. *Clamp connections* not seen.

Since no fresh material of the species has been available to this author, it is very difficult to make a satisfactory disposition of this bolete. The microscopic features clearly indicate that it is a *Suillus*, but because of the peculiar combination of characters its placement in the proper subgeneric categories is beset with problems. The most distinctive features are the almost dry, subtomentose cap (although the notes indicate it was "shiny" when fresh), the yellow context that does not change to blue upon exposure, the large pores and the relatively large spores.

The color of the cap of the type specimen is near cinnamon to ochraceous tawny with slightly darker fibrils, and its appearance is not truly suggestive of any *Suillus* known from the state. *Suillus fuscotomentosus* is much darker colored, has small pores and often an unpleasant taste, but like *S. californicus* the cap surface is strongly fibrillose. Also, *S. fuscotomentosus* has relatively large spores and apparently forms mycorrhizal associations with ponderosa pine. The two species are probably the most closely related of those in this area. *Suillus tomentosus,* another species with a fibrillose pileus, is distinctly viscid, more yellow in color, has fibrils that are often gray when young and reddish when old, and the context changes to blue when exposed.

Edibility unknown.

SUILLUS FUSCOTOMENTOSUS *Thiers and Smith, Contr. Toward a Monogr. of N. Am. Sp. of Suillus, p. 65. 1964*

Illustrations: See Microfiche No. 41

Pileus 3.5–15 cm broad when mature, when young subconic to convex to obtusely convex, becoming broadly convex to plano-convex, occasionally highly irregular in shape with undulating margin; surface moist to subviscid with age, conspicuously fibrillose to fibrillose-scaly to almost squamulose when young, fibrils often becoming agglutinated at the tips, sometimes becoming less dense with age and appearing appressed, occasionally appearing glabrous when very old; color when young olive brown to near fuscous ("clove brown" to "bone brown" to "fuscous") on the disc, unchanging or fading slightly toward the margin, with age slowly fading to dark vinaceous brown ("rood's brown" to "Vandyke brown" to "Verona brown" to "warm sepia"), eventually becoming near cinnamon ("clay color" to "tawny-olive" to "cinnamon" to "sayal brown"), occasionally remaining dark brown ("Verona brown") on the disc and fading to pinkish brown ("cinnamon" to "avellaneous") on the margin, fibrils more or less concolorous or darker than the surface; margin entire, glabrous, no evidence of partial veil during any stage of development. *Context* 1–2 cm thick on the disc, yellow ("ivory yellow" to "cartridge buff") in young pilei, changing to deep yellow ("primrose yellow" to "cream color" to "naphthalene yellow") when older, unchanging when exposed. Taste not distinctive to weakly acid and unpleasant; odor not distinctive.

Tubes up to 1.5 cm long, adnate to shallowly and narrowly depressed around the stipe, unchanging or becoming subdecurrent to broadly shallowly depressed with age; when young pale yellow ("deep colonial buff" to "colonial buff" to "deep olive-buff"), changing to yellow to buff ("antimony yellow" to "warm buff"), finally becoming olive yellow ("Isabella color" to "chamois" to "old gold" to "olive lake"), unchanging when exposed; *pores* up to 1 mm broad, angular, concolorous, unchanging when bruised.

Stipe 4–10 cm long, 1–3 cm thick at apex, equal to typically obscurely clavate to clavate, with white to pale lavender ("avellaneous") mycelium at the base, solid; surface strongly and often noticeably punctate, glandulae relatively large and often becoming elongated with age, more or less concolorous with surface, darkening when handled, moist but not viscid; color when young yellow ("Naples yellow" to "mustard yellow"), typically becoming pallid ("pale olive-buff" to "warm buff"), sometimes becoming bright yellow ("amber yellow" to "strontian yellow") at the apex, occasionally becoming pinkish ("pale pinkish buff") particularly at the base, unchanging upon bruising; no evidence of an annulus. *Context* concolorous with surface, unchanging when exposed.

Spore print olive brown. *Spores* 9–12 X 3–4 μm, smooth, thin-walled, fusoid to subellipsoid to subcylindric in face view, slightly inequilateral in profile, pale yellow in KOH, pale ochraceous in Melzer's.

Basidia 21–24 X 6–8 μm, apparently two- and four-spored, hyaline, clavate. *Hymenial cystidia* 30–45 X 8–10 μm, cylindric to subclavate to obtusely fusoid to highly irregular in outline, in clusters, rare to absent on sides of tubes, abundant on the pores; typically staining dark brown to black in KOH, and appearing incrusted near the pores, often only basal portion of cluster staining when located on side of tubes.

Tube trama hyaline, subgelatinous in KOH, divergent, hyphae ±5 μm wide. *Pileus trama* interwoven, homogeneous. *Pileus cuticle* differentiated as a highly uneven trichodermium of more or less erect, septate hyphal tips, not at all or only slightly gelatinous in KOH, hyphae typically appearing incrusted, and staining pale brown in KOH, cells ±7 μm wide. *Stipe cuticle* not gelatinous in KOH, with numerous large clusters of caulocystidia 60–90 X 6–10 μm, which stain dark vinaceous to black in KOH, cylindric to highly irregular in outline. *Clamp connections* not seen.

Chemical reactions NH_4OH—context gray; HNO_3 and HCl—cuticle yellow, context and tubes lavender; H_2SO_4—cuticle, context, and tubes yellow; $FeSO_4$—context gray.

Habit, habitat, and distribution Solitary to gregarious to occasionally cespitose in soil under ponderosa pine. Originally described from collections made in Santa Cruz County, this species has now been found up to elevations of 3,000 to 4,000 feet in the Sierra foothills and at similar elevations in Castle Crags State Park and vicinity in the Cascade Range. Although often seen in mixed forests, no fruitings have ever been encountered in areas where ponderosa pines were not a conspicuous component of the forest cover. It is often abundant in the fall and winter fruiting season, particularly in the Castle Crags area.

Material studied Butte County: Ripley 1613. Santa Cruz County: Sundberg 178, 211; Thiers 10622, 10623, 10759, 10794, 10981. Shasta County: Thiers 18219, 18220, 18714, 21597, 21622, 21679, 23087. Tuolumne County: Thiers 18618.

Observations This is a very distinctive and easily recognized species. Its most characteristic features are the dark-colored fibrils on the surface of the pileus, the moist to only subviscid cuticle, the unchanging context, and the apparent association with only ponderosa pine. *Suillus fuscotomentosus* appears rather closely related to *S. californicus,* but that species is so poorly known that no definitive comparisons can be made. *Suillus fuscotomentosus* is also somewhat similar in appearance to the dark forms of *S. acerbus,* but that species has a distinctly viscid pileus, the cuticle is differently organized, and it is associated, so far as is known, only with Monterey pine. *Suillus tomentosus,* another species with a conspicuously fibrillose cuticle, is easily distinguished by the much paler color of the pileus and the blue discoloration of the context when bruised or exposed.

This species is not poisonous, but is rated very low as an esculent because of the insipid taste and unpleasing texture of the cooked flesh.

SUILLUS GLANDULOSIPES *Thiers and Smith, Contr. Toward a Monogr. of N. Am. Sp. of Suillus, p. 86. 1964*

Illustrations: See Microfiche No. 42

Pileus 6–12 cm broad at maturity, convex when young, becoming broadly convex to plano-convex to plane or shallowly depressed with age, often highly uneven to undulating or irregular toward the margin when old; surface viscid, when young glabrous or with irregularly distributed fibrils near the margin and more rarely on the disc, typically becoming glabrous with age, with the gluten drying in streaks, rarely appearing somewhat fibrillose and pitted or rugose when older; color at first yellow to pale tan or reddish cinnamon ("pale ochraceous-buff" to "pinkish buff" to "vinaceous-cinnamon" to "orange-cinnamon" to "cinnamon" to "apricot buff" to "ochraceous-orange"), unchanging with age or becoming reddish brown to brown ("ochraceous-tawny" to "Mikado brown" to "Verona brown" to "warm sepia" to "cinnamon-rufous"); margin strongly incurved to inrolled, with a conspicuous white cottony roll when young, disappearing with age, entire. *Context* 1–2 cm thick, white when very young, changing to pale yellow (near "marguerite yellow" to "cartridge buff" to "ivory yellow"), unchanging when bruised or rarely becoming pale lavender ("avellaneous") with age, firm. Taste and odor not distinctive.

Tubes 0.4–1 cm long, decurrent to subdecurrent to occasionally shallowly depressed, when young tan ("warm buff" to "light ochraceous-buff" to "pale ochraceous-buff" to "cream buff"), changing to yellow ("antimony yellow" to "chamois") with age, often with reddish brown glandulae on the pores, unchanging when bruised; *pores* angular, one to two per millimeter, concolorous, unchanging when bruised.

Stipe 4–11 cm long, 1–2 cm thick at the apex, with white mycelium at the base, equal, solid; surface dry, white to pale yellow ("marguerite yellow" to "primrose yellow") to occasionally pale pink ("pale pinkish buff"), unchanging with age, densely covered with irregularly shaped glandulae that are dark brown to black and often stain fingers dark brown when handled; no annulus. *Context* white when young, becoming tan to pale yellow ("warm buff" to "ivory yellow") with age, usually unchanging when exposed, but sometimes darkening slightly at the base, or becoming vinaceous in older basidiocarps.

Spore print brown. *Spores* 6–9(12) × 3–4 μm, hyaline to pale ochraceous in KOH, smooth, thin-walled, cylindric to subellipsoid.

Basidia 23–28 × 5–8 μm, clavate, four-spored, hyaline, contents often granulose in KOH. *Hymenial cystidia* 30–50 × 7–12 μm, abundant only on the pores, fasciculate, incrustations on basal portion staining dark brown when revived in KOH, contents sometimes staining brownish, no solitary cystidia seen.

Tube trama hyaline, divergent from a distinct mediostratum, subgelatinous, hyphae up to 7 μm wide, tissue at junction of pileus and tube trama often staining pale vinaceous in KOH. *Pileus trama* interwoven, homogeneous. *Pileus cuticle* differentiated as a broad (330 μm) ixotrichodermium of interwoven hyphae, contents of hyphae appear brownish-granulose in KOH, outermost hyphae often somewhat enlarged and incrusted, hypodermium well differentiated as a compactly interwoven layer staining dark brown in KOH. *Stipe cuticle* differentiated as a gelatinous layer with numerous fascicles of darkly staining caulocystidia similar to the hymenial cystidia. *Clamp connections* absent.

Chemical reactions unknown.

Habit, habitat, and distribution Gregarious to cespitose in humus under Bishop and, less commonly, beach pine. In California, *Suillus glandulosipes* is known only from the northern coastal regions. It is relatively rare in California and, with the exception of one collection made in Patricks Point State Park in Humboldt County, it has been found only in Jackson State Forest in Mendocino County. Although originally described from California, it is now known from throughout the Pacific Northwest and as far east as Michigan.

Material studied Humboldt County: Thiers 14516. Mendocino County: Thiers 8394, 9307, 9332, 9335, 9472, 10702, 14150, 18106, 21414, 21443, 23063.

Observations *Suillus glandulosipes* is distinguished by the pale cinnamon pink to more vinaceous colored pileus, the large conspicuous glandulae on the stipe, and the well-developed, easily detected ring of velar tissue on the margin of the pileus. No evidence of any type of annulus has ever been observed. The ring of veil tissue is readily apparent on young basidiocarps but characteristically disappears with age. This species obviously belongs to the same group as *S. granulatus* and *S. albidipes.* The presence of the veil tissue in *S. glandulosipes* is perhaps the most significant difference, but the pale color of the pileus is also quite distinctive. Although it is indicated in the description that old basidiocarps may darken considerably, it should be emphasized that this is not the usual pattern. *Suillus brevipes* is commonly found in the same type of habitat and often fruits at the same time, but it characteristically has a much shorter stipe and lacks the noticeable glands on the surface.

Edibility not determined.

SUILLUS GRANULATUS *(Fries) Kuntze, Rev. Gen. Pl. 32:535. 1898*

Boletus granulatus Fries, Syst. Mycol. 1:385. 1821.

Rostkovites granulatus (Fries) Karsten, Rev. Mycol. 3:16. 1881.

Illustrations

Kawamura, S., Icones of Japanese Fungi, pl. 232.

Lange, J. E., and M. Lange, 600 Pilze in Farben, p. 189.

Leclair, A., and H. Essette, Les Bolets, pl. 9.

Romagnesi, H., Nouvel Atlas des Champignons, pl. 120A.

Singer, R., Die Rohrlinge, Teil I, pl. VII, figs. 9–13, pl. VIII, figs. 1–10.

Smith, A. H., The Mushroom Hunter's Field Guide, p. 90, pl. 56.

Smith, A. H., and H. D. Thiers, A Contribution Toward a Monograph of North American Species of Suillus, pls. 43, 44.

Smith, A. H., and H. D. Thiers, The Boletes of Michigan, pl. 32.

Snell, W. H., and E. A. Dick, The Boleti of Northeastern North America, pl. 20.

Pileus 5–12 cm broad when mature, convex, becoming broadly convex to plano-convex to plane; surface viscid to glutinous, glabrous or streaked or spotted with brown ("cinnamon") on a pale buff ("pinkish buff" to "cinnamon-buff") ground color, in age more or less cinnamon ("orange-cinnamon") overall, immature specimens often pallid for a long time; margin naked to minutely tomentose, in age often areolate, incurved, becoming decurved, no conspicuous cottony roll of veil tissue. *Context* 1–2 cm thick, whitish when young, unchanging or becoming pale yellow, with a watery green line above the tubes. Taste mild to acid; odor mild to slightly fragrant.

Tubes up to 1 cm long, adnate to subdecurrent, pallid at first, but soon pale yellow, becoming dingy yellow, not staining when bruised; *pores* small, about two per millimeter, or in old basidiocarps one per millimeter and somewhat boletinoid to merely elongated, pale honey yellowish, in age brownish spotted, staining dingy cinnamon when bruised, in very young stages beaded with droplets of a cloudy liquid.

Stipe 4–8 cm long, 1–2 cm thick at the apex, equal or narrowed to a point at the base, solid; surface whitish, but soon bright yellow above, pallid downward, base dingy cinnamon, covered overall by pinkish-tan to vinaceous-brown glandular dots; no annulus. *Context* white inside at first but soon bright yellow within at apex, tinged cinnamon toward base.

Spore print dingy cinnamon to cinnamon. *Spores* 7–9 X 2.5–3.5 μm, smooth, yellowish in KOH and Melzer's, subellipsoid, in side view somewhat inequilateral.

Basidia 18–24 X 5–6 μm, four-spored, clavate, hyaline in KOH, merely yellowish in Melzer's. *Hymenial cystidia* 36–50 X 7–9 μm, clavate to subcylindric, hyaline or with yellowish to brown content, in clusters with brown incrusted material around the base.

Tube trama of gelatinous divergent hyphae from a colored central strand. *Pileus trama* interwoven, homogeneous. *Pileus cuticle* an ixotrichodermium composed of hyphae 4–7 μm wide, in KOH the contents dingy ochraceous to bister, some incrusting material present in the layer. *Caulocystidia* 40–70 X 7–10 μm, clavate, subfusoid to cylindric, bases of clusters surrounded by brown pigment, contents hyaline to dark brown. *Clamp connections* absent.

Chemical reactions unknown.

Habit, habitat, and distribution Gregarious to scattered in humus under pines. This is truly a cosmopolitan species and has been found in most of the pine forests of the United States. It is common in the coastal forests of northern California and also occurs in the Cascade and Sierra Nevada ranges.

Material studied Madera County: Thiers 20859, 20955. Mendocino County: Thiers 8166, 8204, 8301, 8775, 8789, 8962, 8999, 9263, 9514, 9633, 10601, 11078. Mono County: Thiers 21000. Nevada County: Thiers 13157, 13249. Shasta County: Thiers 21595. Tuolumne County: Thiers 21066, 21092.

Observations This very common and well-known *Suillus* needs little comment. It is distinguished by the reddish brown color of the pileus, the absence of an annulus or even a ring of veil tissue on the margin of the pileus, the presence of conspicuous glands on the surface of the stipe, and a mild, pleasant taste. The absence of any type of veil is the major feature separating this species from both *S. albidipes* and *S. glandulosipes.* In the San Francisco Bay Area it has been confused with *S. pungens,* which, however, has a harsh, unpleasant taste and is typically white when young, becoming gray and finally rusty brown. Furthermore, young basidiocarps of *S. pungens* have a distinct cottony roll of veil tissue. Pantidou, in comparing pure cultures of the two species, has likewise found them to be highly distinct.

Edible and commonly eaten in many areas of the United States.

SUILLUS LUTEUS *(Fries) S. F. Gray, Nat. Arr. Brit. Pls. 1:646. 1821*

Boletus luteus Fries, Syst. Mycol. 1:385. 1821.

Illustrations

Kawamura, S., Icones of Japanese Fungi, pl. 229.
Lange, J. E., and M. Lange, 600 Pilze in Farben, p. 187.
Leclair, A., and H. Essette, Les Bolets, pl. 8.
Romagnesi, H., Nouvel Atlas des Champignons, pl. 120B.
Singer, R., Die Rohrlinge, Teil I, pl. V, figs. 13–21.
Smith, A. H., The Mushroom Hunter's Field Guide, p. 82, pl. 49.
Smith, A. H., and H. D. Thiers, A Contribution Toward a Monograph of North American Species of Suillus, pl. 23.
Smith, A. H., and H. D. Thiers, The Boletes of Michigan, pl. 17.
Snell, W. H., and E. A. Dick, The Boleti of Northeastern North America, pl. 17

Pileus 5–12 cm broad, hemispheric to obtuse at first, broadly convex to plane in age; surface glabrous and viscid, but at times somewhat streaked beneath the gluten; color yellow brown to red brown ("raw sienna" to "Sudan brown" or "tawny" to "bister"), pellicle separable. *Context* white or tinted more or less with pale yellow, especially near tubes and stipe. Taste pleasant; odor not distinctive.

Tubes 3–7 mm deep, gradually shorter toward the cap margin, adnate to subdecurrent, at first whitish to pale yellow, at length dark yellow ("honey yellow" to "old gold"), unchanging when cut; *pores* three per millimeter, in age one to two per millimeter, yellow at first, but becoming dotted with dark glandulae.

Stipe (3)4–8 cm long, 1–2.5 cm thick, equal or attenuate at base, solid, typically peronate by the whitish veil up to the membranous, often reflexed, rather persistent annulus, glandular dotted above the annulus and pale yellow, glandular dotted beneath the sheath; veil membranous and with a gelatinous, purplish zone on the underside, becoming purplish gray in age. *Context* concolorous with surface or whitish.

Spore print brown, dull cinnamon when moisture has escaped. *Spores* 7–9 X 2.5–3 μm, smooth, with a pronounced hyaline sheath, more or less oblong in face view, narrowly inequilateral in profile view, nearly hyaline in KOH, yellowish in Melzer's.

Basidia 14–18 X 4–5 μm, four-spored, yellowish in Melzer's and KOH. *Hymenial cystidia* in scattered to rare bundles surrounded by bister debris (details of bundle obscured by overall covering of debris), when isolated 20–35 X 5–7 μm and narrowly clavate, content usually bister as revived in KOH.

Tube trama gelatinous and divergent. *Pileus trama* of interwoven, hyaline, nonamyloid hyphae. *Pileus cuticle* of narrow (2–4 μm), gelatinous filaments forming a tangled trichodermium, hyaline in KOH and pale yellowish in Melzer's. *Stipe cuticle* with numerous fascicles of caulocystidia, bister in KOH. *Clamp connections* not seen.

Chemical reactions unknown.

Habit, habitat, and distribution This species has not been collected in California by the author, and the only indication of its presence in the state is from reports that were published around the turn of the century. It is, however, a common and widely distributed species in North America, and its presence in California would not be very surprising. It enjoys a rather broad range of apparent mycorrhizal hosts and has been commonly found under both pine and spruce.

Observations *Suillus luteus* is characterized by a very viscid to gelatinous pileus, a conspicuous, often flaring and viscid annulus that typically has a lavender band or layer on the external surface, and prominent glandulae on the stipe during all stages of development. As has been pointed out by Smith and Thiers, this species is affected by various environmental factors and, therefore, often appears highly variable. The partial veil and annulus, for example, seem to be influenced considerably by the amount of moisture in the air. If the air is dry, the annulus is likewise dry and the lavender layer may not be apparent; on the other hand, if the humidity is high, the annulus is viscid to glutinous and the band easily detected. In some basidiocarps the velar tissue may remain attached to the margin of the pileus. In such condition it is likely to be confused with *S. albidipes* or even *S. granulatus*.

Edible.

SUILLUS MEGAPORINUS *Snell and Dick, Mycologia 48: 302. 1956*

Illustrations: See Microfiche No. 44

Pileus 2–5 cm broad at maturity, globose to convex when young, becoming broadly convex to plano-convex to highly irregular with age; surface subviscid to viscid, streaked to fibrillose to appressed fibrillose scaly, scales occasionally becoming free when very old; background color tan ("warm buff" to "ochraceous-buff"), scales near light brown ("cinnamon-buff" to "avellaneous"), sometimes becoming rust colored ("ochraceous tawny") when older; margin incurved, with obscure partial veil fragments when young, glabrous at maturity. *Context* 0.5–1 cm thick, tan ("warm buff" to "ochraceous-buff" to near "cinnamon-buff"), unchanging when exposed. Taste and odor not distinctive.

Tubes up to 1 cm long, depressed, becoming arcuate-decurrent to strongly decurrent with age, yellow ("antimony yellow" to "chamois"), sometimes reddish ("tawny" to "russet") tints with age, unchanging when exposed, often appearing similar in arrangement to that of *Gastroboletus turbinatus; pores* up to 5 mm in length and 3 mm broad, intervenose, almost lamellate, with thick, wrinkled, irregular margins, concolorous with tubes, unchanging when bruised.

Stipe 1–3 cm long, 4–7 mm thick at apex, often appearing short and reduced, sometimes somewhat eccentric, pallid to more or less concolorous with the pileus; surface becoming conspicuously glandular dotted at maturity; no annulus, but obscure fibrillose annular zone sometimes apparent. *Context* concolorous with that of the pileus, unchanging.

Spore print color unknown. *Spores* 8.5–10.5 X 4.5–6 μm, ellipsoid to subfusoid to subcylindric, thin-walled, hyaline in KOH, pale yellow in Melzer's.

Basidia 25–30 X 5–8 μm, hyaline in KOH, cylindric, four-spored. *Hymenial cystidia* 30–48 X 7–11 μm, rare on sides of tubes, abundant to crowded on pores, fascicled, no solitary cystidia seen, dark brown in KOH, chocolate brown in Melzer's, cylindric to clavate, thin-walled, heavily incrusted.

Tube trama hyaline, appearing subgelatinous in KOH, parallel to regular to obscurely divergent, hyphae 3–4 μm wide. *Pileus trama* loosely interwoven, homogeneous, hyaline to pale brown in KOH. *Pileus cuticle* differentiated as an ixotrichodermium, hyaline to pale buff to pale brown in KOH, rust red ("ochraceous-tawny") in Melzer's; hypodermium compactly interwoven, dark brown in KOH and Melzer's. *Stipe cuticle* gelatinous in KOH, interwoven, with large clusters of caulocystidia staining dark brown in KOH, often with strong lavender tints, similar in shape and size to the hymenial cystidia. *Clamp connections* absent.

Chemical reactions NH_4OH–cuticle grayish lavender, context rose color, tubes red; HNO_3, HCl, H_2SO_4–cuticle orange, context deep lilac, tubes red; $FeSO_4$–cuticle blue gray.

Habit, habitat, and distribution Gregarious to scattered in soil under conifers at edge of seepage areas. Basidiocarps often occurring in large numbers.

Material studied Fresno County: Thiers 13423, 13440. Tuolumne County: Thiers 21051, 21084, 21085, 21087, 21095, 21097, 21105, 21107.

Observations At the time the monograph by Smith and Thiers was published, no fresh material of this species had been seen. Since that time, however, collections have been made in the type locality and in other areas in the Sierra. This is a highly distinctive *Suillus.* The arrangement and disposition of the tubes cause the basidiocarp to appear diseased or gastroid, but a spore deposit is delivered. The pores are larger than in any other *Suillus* and are so long radially that the hymenophore appears distinctly lamellate. The stipe is very reduced, short, and somewhat eccentric. Like *S. americanus* it sometimes has reddish streaks or fibrils on the surface, but does not resemble that species otherwise. In Snell and Dick's original description, they noted that an ample annulus was sometimes present. This was not apparent on any of the basidiocarps seen, although a fibrillose annular zone could be detected on some. *Suillus megaporinus* may be somewhat closely related to *S. riparius* since both species have rather large pores. However, *S. riparius* has a well-developed stipe, a much darker-colored pileus, with a rough scaly surface, larger cystidia, and a palisade type of cuticle on the stipe.

Edibility unknown.

SUILLUS MONTICOLUS *Thiers, Madrono 19:154. 1967*

Pileus 4–7 cm broad when mature, convex when young, becoming plane to plano-convex to broadly convex to highly irregular with age; surface viscid to subviscid, streaked to fibrillose to appressed fibrillose-scaly; color when young light brown ("clay color" to "pale olive-buff" to "olive-buff"), with age changing to brown ("buffy brown" to dark "avellaneous" to near "cinnamon-brown"), more or less evenly colored although the streaks or fibrils may be slightly darker; margin incurved, becoming decurved with age, glabrous during all stages of development with no evidence of a partial or false veil. *Context* 0.5–1 cm thick, white, unchanging when exposed. Taste and odor not distinctive.

Tubes 2–5 mm long, adnate to shallowly depressed, becoming decurrent to subdecurrent with age, strongly radiately arranged, when young bright yellow ("straw yellow" to "amber yellow"), darkening with age to dark yellow ("antimony yellow" to "old gold"); *pores* 1–2 mm broad, often compound, concolorous with the tubes, unchanging when bruised, often staining vinaceous in KOH.

Stipe 3–5 cm long, 2–4 cm thick at the apex, sometimes eccentric, distinctly ventricose to bulbous at the base and abruptly pinched below, usually bent, sometimes almost at right angles, white mycellium at base, stuffed; surface dry, glabrous when young, becoming glandulose to punctate to occasionally obscurely fibrillose with age, glandulae apparent only in the apical region, frequently reticulate at the apex, concolorous with the tubes at the apex, usually becoming brown ("olive-buff" to "cinnamon-brown") toward the base; no evidence of an annulus or partial veil. *Context* white, unchanging when exposed.

Spore print color unknown. *Spores* 8.5–11.2 X 3–4 µm, cylindric to subellipsoid, hyaline in KOH, pale yellow in Melzer's, smooth, thin-walled.

Basidia 30–35 X 7–10 µm, clavate to subcylindric, two- and four-spored, hyaline in KOH. *Hymenial cystidia* 25–45 X 4–9 µm, cylindric to subclavate, clustered, often massed, particularly abundant on the pores, scattered to numerous on the sides of the tubes, staining dark brown in KOH and Melzer's, usually staining more darkly on the pores; no solitary cystidia seen.

Tube trama divergent from a distinct mediostratum, hyaline in KOH, appearing gelatinous to subgelatinous in KOH, hyphae 3–6 µm wide. *Pileus trama* loosely interwoven, homogeneous, more or less hyaline in KOH. *Pileus cuticle* clearly differentiated as a trichodermium of more or less upright to somewhat tangled hyphae with free tips, walls appearing finely incrusted, subgelatinous in KOH, staining chocolate brown in KOH, ochraceous in Melzer's; hypodermium compact, closely interwoven, staining brown to ochraceous in KOH, ochraceous tawny in Melzer's. *Stipe cuticle* hymeniform, with scattered clusters of caulocystidia similar to the cystidia on the pores. *Clamp connections* not seen.

Chemical reactions unknown.

Habit, habitat, and distribution Gregarious in soil under lodgepole pine. Elevation 7,200 ft.

Material studied El Dorado County: 32764. Nevada County: Thiers 13248.

Observations The most distinctive feature of *Suillus monticolus* is the highly bulbous to ventricose stipe that is frequently bent and often somewhat eccentric. The glandulae, while always present, are never strongly developed. The hymeniform cuticle of the stipe also appears highly distinctive.

Edibility unknown.

SUILLUS PUNCTATIPES *(Snell and Dick) Smith and Thiers, Contr. Toward a Monogr. of N. Am. Sp. of Suillus, p. 94. 1964*

Boletinus punctatipes Snell and Dick, Mycologia 33: 36. 1941.

Illustrations

Smith, A. H., and H. D. Thiers, A Contribution Toward a Monograph of North American Species of Suillus, pl. 45.

Smith, A. H., and H. D. Thiers, The Boletes of Michigan, pls. 33, 34.

Pileus 6–16 cm broad, convex when young, becoming broadly convex to highly irregular in outline; surface very viscid, glabrous but often appearing streaked by the gluten, sometimes appearing appressed fibrillose on the margin, smooth; color when young brown ("walnut brown" to "burnt umber" to "Vandyke brown") to sometimes dark brown ("warm sepia" to "bister"), unchanging or fading to light brown ("cinnamon" to "orange-cinnamon" to "Mikado brown") with age; margin strongly incurved when young, becoming decurved to plane to irregular with age, entire. *Context* 1–2 cm thick, white, unchanging when exposed. Taste mild; odor pungent.

Tubes 1 cm long, decurrent, occasionally slightly depressed, whitish to tan ("ochraceous-buff") when young, changing to yellow ("antimony yellow" to "yellow ocher"), unchanging or becoming brownish when bruised, strongly boletinoid during all stages of development; *pores* 1–3 mm long, compound, concolorous.

Stipe 3–5 cm long, 1–2 cm thick at the apex, equal to somewhat pinched at the base, solid; surface strongly glandulose, yellow at the apex, whitish below, darkening when handled, dry; no annulus. *Context* white, occasionally changing to yellow, typically vinaceous at the base.

Spore print brown. *Spores* 7.5–9 X 3–4 μm, subellipsoid to oblong, ventricose in side view, smooth, thin-walled, hyaline in KOH and Melzer's.

Basidia 27–32 X 6–9 μm, clavate, four-spored. *Hymenial cystidia* 45–60 X 7–10 μm, scattered to numerous, clustered, cylindric to clavate, hyaline, incrusted, dark brown in KOH.

Tube trama divergent, gelatinous, hyphae 4–6 μm wide. *Pileus trama* interwoven, homogeneous. *Pileus cuticle* differentiated as an ixotrichodermium of tangled hyphae, incrusted. *Stipe cuticle* with caulocystidia similar to the hymenial cystidia. *Clamp connections* absent.

Chemical reactions unknown.

Habit, habitat, and distribution Solitary to gregarious in soil under conifers. A single collection of this species has been made in the vicinity of Donner

Pass in Nevada County. It is most probably associated with lodgepole pine, but the basidiocarps were collected in a mixed conifer forest. The pines, however, were dominant, and the basidiocarps were closer to them than to other conifers.

Material studied Nevada County: Sundberg 351.

Observations This *Suillus,* originally described from collections made in the state of Washington, rather closely resembles *Suillus granulatus* but is readily distinguished by the noticeable boletinoid arrangement of the tubes and pores, which occur in distinct rows radiating from the center of the pileus. The white context that, so far as is known, does not change to yellow with age is another distinctive feature. Some basidiocarps of *S. punctatipes* have the stature of *S. brevipes,* but their stipes are usually noticeably punctate, at least in the upper portion.

Edibility unknown.

SUILLUS PUNGENS *Thiers and Smith, Contr. Toward a Monogr. of N. Am. Sp. of Suillus, p. 92. 1964*

Illustrations: See Microfiche No. 47

Pileus 4–14 cm broad when expanded, obtusely convex to convex when young, becoming broadly convex to plano-convex with age; surface viscid to glutinous during all stages of development, glabrous but sometimes appearing obscurely streaked from gluten when older; color when young white to olive ("deep grayish olive" to "grayish olive" to "citrine drab") with splotches or irregularly shaped areas colored pale olive ("pale olive-buff" to "olive-buff"), frequently strongly variegated with a mixture of light and dark colors, when older unchanging or becoming reddish brown ("ochraceous-tawny"), sometimes a mixture of all the pigments mentioned above; margin incurved and with a cottony roll of white tissue when young, becoming naked and merely decurved with age. *Context* 1–2 cm thick, white and unchanging in young basidiocarps, frequently changing to yellow ("pinard yellow") when older. Taste harsh, subnauseous, and weakly acid; odor strong, pungent.

Tubes up to 1 cm in length, adnate when young, becoming decurrent to subdecurrent with age; color when young whitish to pale buff ("cartridge buff") with conspicuous whitish droplets that become brown to ochraceous when dried, with age changing to near yellow ("colonial buff") and finally to dark yellow ("honey yellow"); *pores* 1–1.5 mm broad, not radially arranged, unchanging when bruised, angular.

Stipe 3–7 cm long, 1–2 cm thick at the apex, equal to tapering at the base to sometimes subventricose, solid; surface dry, glabrous, strongly punctate, glandulae large, irregular in outline, reddish at first then becoming brownish, background whitish to more or less concolorous with the tubes when young, becoming yellow ("pinard yellow" to "massicot yellow") with age, unchanging when bruised; no annulus. *Context* white, unchanging when exposed.

Spore print brown. *Spores* 9.5–10 X 2.8–3.5 μm, hyaline in KOH, smooth, thin-walled, ellipsoid to subcylindric in face view, inequilateral in profile.

Basidia 33–36 X 8–10 μm, hyaline, clavate, contents granulose in KOH, four-spored. *Hymenial cystidia* 43–79 X 7–10 μm, rare to scattered, abundant on the pores, typically occurring in massive clusters, dark brown in KOH, cylindric to subclavate, incrusted, occasionally hyaline, thin-walled.

Tube trama hyaline, divergent to subparallel, hyphae 3–5 μm wide. *Pileus trama* interwoven, homogeneous. *Pileus cuticle* differentiated as an ixotrichodermium, staining brown in KOH, hyphae 4–5 μm wide. *Stipe cuticle* with clusters of cystidia similar to those found in the hymenium. *Clamp connections* absent.

Chemical reactions KOH–context vinaceous, tubes red, pileus cuticle black, stipe cuticle pale vinaceous; NH_4OH–context very pale vinaceous, tubes bright red; $FeSO_4$–context gray, tubes dark gray to black, stipe cuticle light gray.

Habit, habitat, and distribution Solitary to gregarious in humus under Monterey pine. This is often the most abundant *Suillus* in the San Francisco Bay Area, and although it is usually found under Monterey pine, a few collections have been made under knobcone pine. The type collection was made on the campus of San Francisco State University in San Francisco, where it occurs in abundance during the fall and winter seasons. It is one of the few species of *Suillus* that continues to fruit sporadically throughout the year.

Material studied Marin County: Ripley 938; Thiers 7376. San Francisco County: Johnson 1; Largent 55, 143, 317; Peters 36, 49, 239, 271, 588; Thiers 8570, 9330, 11772, 11804, 17310, 21393, 21908, 27515. San Mateo County: Thiers 7505, 11912, 18178. Santa Cruz County: Thiers 10748. Sonoma County: Sundberg 72.

Observations This species is distinctively characterized by the very noticeable color changes occurring in the pileus as it develops. In most instances the young pilei are pure white, as development continues they next become dark gray, and finally at maturity they typically change to rust brown in color. The tubes, when young, are almost white but become yellow with age. If the humidity is high, there is usually an abundance of whitish droplets on the hymenophore, at least during the young stages. The ring of cottony veil tissue on the margin of the pileus is very pronounced during the younger stages but, as is often the case in the genus *Suillus*, it disappears with age. *Suillus pungens* has been misidentified as *S. placidus* because of the white color of the young basidiocarps and the droplets of exudate. The pallid to white colors, the cottony roll of veil tissue, and the unpleasant taste and odor distinguish it from *S. granulatus* and *S. albidipes.*

Edible, but not choice.

SUILLUS RETICULATUS *sp. nov.*

Pileus 6–8 cm latus, subviscidus vel viscidus, glaber vel appresso-fibrillosus, brunneus vel rufobrunneus. Contextus 9–14 mm crassus, flavus tactu caeruleus. Sapor et odor mitis. Tubuli 0.5–1 cm longi, olivaceo-brunnei tactu caerulei; pori 1 mm lati, concolores tactu caerulei. Stipes 10–12 cm longus, 1–1.5 cm crassus, reticulatus, siccus, flavus; contextus flavus tactu caeruleus. Sporae 9–11 X 3–4 μm, subfusoideae vel subellipsoideae. Cystidia 18–35 X 7–10 μm, fasciculata. Cuticula viscida, intertexta. Holotypus (no. 23616) a H. D. Thiers lectus prope Agnew Meadows, Madera County, August 18, 1969; in Herbarium San Francisco State University conservatus.

Pileus 6–8 cm broad when mature, convex when young, becoming plano-convex with a flared or uplifted margin when mature; surface subviscid to viscid, glabrous to obscurely appressed fibrillose during all stages of development; color brown to reddish brown ("tawny" to "ochraceous-tawny" to "buckthorn brown"); margin entire, glabrous, eroded when old. *Context* 9–14 mm thick, yellow, changing to blue when exposed. Taste and odor mild.

Tubes 0.5–1 cm long, adnate to adnexed, olive brown ("olive-brown"), changing to blue when bruised or exposed; *pores* up to 1 mm broad, angular, concolorous with the tubes, bluing when bruised.

Stipe 10–12 cm long, 1–1.5 cm thick at the apex, subequal to subclavate, solid; surface distinctly reticulate on upper one half to two thirds portion, dry, yellow ("antimony yellow") at the apex, darkening slightly toward the base, reticulum concolorous; no evidence of a partial veil. *Context* yellow in the apex, becoming vinaceous toward the base, bluing in the apex when exposed.

Spore print color unknown. *Spores* 9–11 X 3–4 μm, pale ochraceous in KOH and Melzer's, subfusoid to subellipsoid in face view, inequilateral in profile, walls thin, smooth.

Basidia 20–25 X 7–9 μm, four-spored, clavate, hyaline in KOH. *Hymenial cystidia* 18–35 X 7–10 μm, in large clusters or fascicles, basal portion of cluster black in KOH, apical portion dark brown in KOH, clavate.

Tube trama subgelatinous, hyaline in KOH, divergent, hyphae 5–8 μm wide. *Pileus trama* homogeneous, interwoven. *Pileus cuticle* a closely interwoven ixotrichodermium, pale ochraceous in KOH; hypodermium tightly interwoven, compact. *Stipe cuticle* differentiated as a layer of fertile basidia with fascicles of caulocystidia interspersed. *Clamp connections* absent.

Chemical reactions unknown.

Habit, habitat, and distribution Solitary in soil under lodgepole pine. Known only from the vicinity of the Devil's Post Pile National Monument.

Material studied Madera County: Thiers 23616 type.

Observations This is a very distinct species readily recognized by the combination of an almost glabrous pileus, context and tubes that stain blue, and, most especially, by the noticeably reticulate stipe. No other *Suillus* species known to the author possesses such a stipe. This species does not really have the aspect of a *Suillus* because of the long, reticulate stipe, which lacks prominent glands on the surface. Because of the bluish discoloration of the flesh and the absence of a veil, it appears to be most closely related to *S. tomentosus;* however, the latter has a strongly fibrillose scaly pileus and a more or less smooth stipe. Among the other species in which the context changes to blue, *S. subvariegatus* and *S. ruber* are easily distinguished by either the fibrillose scales on the pileus or distinctive colors of the basidiocarps; *S. caerulescens* and *S. ponderosus* belong to the section *Boletinus* and are not suggestive of this species.

Edibility unknown.

SUILLUS RIPARIUS *Thiers, Madrono 19:156. 1967*

Illustrations: See Microfiche No. 48

Pileus 5–12 cm broad at maturity, convex to obtusely conic when young, becoming plano-convex to plane to broadly convex to shallowly depressed with age, often highly irregular and undulating; surface viscid, when young smooth to glabrous to subglabrous, occasionally somewhat streaked to appressed fibrillose, typically becoming somewhat rimose to areolate to fibrillose-scaly, occasionally in very old pilei appearing somewhat squarrose scaly; color when young brown ("buckthorn brown" to "Dresden brown") to occasionally dark brown ("Prout's brown"), margin yellow ("antimony yellow" to "yellow ocher"), scales brown ("Prout's brown" to "cinnamon-brown") to occasionally rust brown ("ochraceous-tawny"), with age pileus dark yellow ("yellow ocher" to "old gold" to sometimes "Naples yellow" to "straw yellow") to occasionally reddish brown ("ochraceous tawny" to "ochraceous-buff") when very old, scales typically unchanging or sometimes becoming pale vinaceous when bruised; margin when young with a pronounced, large roll of veil tissue that fades to whitish ("pale olive-buff") with age, either appendiculate or glabrous to eroded with age. *Context* up to 1 cm thick, soft, yellow ("light ochraceous-buff" to "massicot yellow" to "Naples yellow"), changing to pinkish ("avellaneous" to "drab") after exposure. Taste acid; odor not distinctive.

Tubes up to 1.5 cm in length, bluntly adnate when young, becoming arcuate-decurrent to decurrent with age; when young deep yellow ("mustard yellow" to "amber yellow"), changing to dark yellow ("antimony yellow" to "yellow ocher") with age, sometimes becoming brownish with age, unchanging upon exposure; *pores* up to 4 mm in length, 3 mm in width, distinctly compound, concolorous with tubes, unchanging when bruised.

Stipe 3–8 cm long, 0.5–1.5 cm thick at the apex, more or less equal, stuffed to hollow; surface moist to subviscid, glabrous except for strongly developed, elongated, irregularly shaped, brown ("Sudan brown" to "amber brown") glandulae; color yellow, often darkening slightly toward the base, frequently with a slight vinaceous tint at the base; no annulus, but fibrillose zone sometimes apparent. *Context* concolorous with flesh of pileus, unchanging when exposed.

Spore print brown. *Spores* 8–11.5 X 3.2–5 μm, subcylindric to subfusoid, hyaline to pale ochraceous in KOH, pale ochraceous in Melzer's, smooth, thin-walled.

Basidia 21–29 × 8–10 μm, clavate to subcylindric, two- and four-spored, hyaline in KOH. *Hymenial cystidia* 35–75 × 5–10 μm, clustered, scattered to numerous, more abundant on the pores, staining dark brown in KOH and Melzer's, sometimes discolored only in base, incrusted, incrustations often floating free in mounting medium, cylindric to subclavate to subfusoid, thin-walled, occasionally hyaline in KOH.

Tube trama hyaline in KOH, divergent from an indistinct mediostratum, subgelatinous in KOH, hyphae 3–5 μm wide. *Pileus trama* interwoven, homogeneous, hyaline in KOH. *Pileus cuticle* differentiated as an ixotrichodermium, staining brown in KOH and ± reddish brown ("ochraceous-tawny") in Melzer's, incrusted, hyphae 3–4 μm wide; hypodermium compactly interwoven, well differentiated, brown in KOH, reddish brown ("ochraceous-tawny") in Melzer's. *Stipe cuticle* hymeniform with large clusters of fascicled caulocystidia, dark reddish brown in KOH. *Clamp connections* not seen.

Chemical reactions unknown.

Habit, habitat, and distribution Gregarious to cespitose in the vicinity of dead logs and stumps near the edges of streams in the vicinity of ponderosa and sugar pine. It is usually found in large numbers during September and October in the Sierra Nevada at elevations of 5,000 to 7,000 feet.

Material studied Calaveras County: Thiers 21253, 21257, 21261. Fresno County: Thiers 13350, 13351, 13356. Tuolumne County: Thiers 13283, 13284, 13883, 13885, 13889, 21294, 21311.

Observations The reddish brown scales on the pileus and the general stature of the basidiocarps of *S. riparius* make it somewhat suggestive of *S. americanus*. The sturdier, generally broader stipe of *S. riparius* in addition to its yellow-brown pores, darker-brown pileus and dark-brown glandulae on the stipe readily distinguish the two species. Also, so far as is definitely known, *S. americanus*. The sturdier, generally broader stipe of *S. riparius* in addition to its presence in the western United States is open to question. The earlier report by Thiers of the presence of *S. americanus* in California is now believed to be in error, and the basidiocarps so identified belong to *S. riparius.* Because of its large, rather labyrinthine pores, *S. riparius* might also be confused with *S. megaporinus,* which, however, has a much paler colored pileus, a poorly developed, often eccentric stipe, narrower spores, and a different type of habitat.

Edibility not determined.

SUILLUS SIBIRICUS *(Singer) Singer, Farlowia 2:260. 1945*

Ixocomus sibiricus Singer, Rev. Mycol. 3:46. 1938.

Boletus sibiricus (Singer) Smith, Mushrooms in Their Natural Habitats, p. 220. 1949.

Illustrations

Singer, R., Die Rohrlinge, Teil I, pl. V, figs. 2–4.

Smith, A. H., Mushrooms in Their Natural Habitats, Reel XXXI.

Smith, A. H., and H. D. Thiers, A Contribution Toward a Monograph of North American Species of Suillus, pls. 34, 35.

Smith, A. H., and H. D. Thiers, The Boletes of Michigan, pl. 24.

Snell, W. H., and E. A. Dick, The Boleti of Northeastern North America, pl. 15.

Pileus 5–10 cm broad when mature, broadly conic to convex when young, becoming broadly convex to plane to umbonate at maturity; surface viscid, when very young appearing glabrous to streaked with gluten, sometimes with plaques or scales toward the margin, with age glabrous on the disc or with scattered inconspicuous plaques to conspicuous plaques on the disc; when young background color yellow ("mustard yellow" to "Isabella color") with the gluten and plaques colored reddish ("russet"), with age background changing to dark yellow ("chamois" to "honey yellow") and the plaques brown to reddish brown ("Mikado brown" to "pecan brown"), tending to stain vinaceous when bruised; margin strongly incurved to inrolled, when young with a highly conspicuous roll of cottony veil tissue colored whitish to pale yellow, but with age whitish to grayish white, becoming glabrous with age. *Context* 1–2 cm thick, yellow ("Naples yellow"), but typically staining vinaceous when exposed. Taste of gluten acid; otherwise taste and odor mild.

Tubes 7–10 mm long, decurrent, when very young yellow ("mustard yellow"), darkening to dull yellow ("antimony yellow" to "yellow ocher") with age, staining vinaceous to dirty pink when bruised; *pores* up to 2 mm broad, elongated to rectangular, typically radially arranged but not always so, compound, concolorous with tubes.

Stipe 5–11 cm long, 0.5–1 cm thick at the apex, equal, solid, sometimes appearing hollow with age; surface moist to subviscid, staining fingers black when handled, glandular, glands pallid and inconspicuous when young, becoming brown to yellow brown with age; background color yellow ("Naples yellow" to "mustard yellow"), staining brown to vinaceous when bruised; annulus sometimes present, often evanescent.

Spore print brown. *Spores* 8–12 X 3.8–4.5 μm, subellipsoid to subcylindric, ventricose in side view, smooth, thin-walled.

Basidia 22–34 X 5–8 μm, clavate, four-spored. *Hymenial cystidia* 50–75 X 6–9 μm, clustered, hyaline but staining dark brown in KOH, scattered to abundant, subclavate to cylindric.

Tube trama divergent, gelatinous in KOH, hyaline. *Pileus trama* interwoven, homogeneous. *Pileus cuticle* differentiated as an ixotrichodermium, hyphae 3–6 μm wide. *Caulocystidia* similar to the cystidia in the hymenium. *Clamp connections* absent.

Chemical reactions KOH—context flashing red then blackening; $FeSO_4$—slowly grayish on cuticle and context.

Habit, habitat, and distribution Scattered to gregarious in soil under western white pine. This species has not yet been recorded from California so far as is known. It is common, however, in several western states, and is included here since western white pine is relatively widely distributed within the northern half of the state.

Observations Earlier reports by Smith et al. have pointed out the apparent confusion between this species, *S. umbonatus,* and *S. americanus.* There are many close similarities between *S. sibiricus* and *S. umbonatus* and the two are frequently misdetermined. *Suillus umbonatus* generally has larger pores, a somewhat smaller, more umbonate pileus that is often pallid in color, and is associated with lodgepole rather than five-needle pine. It is readily apparent that *S. americanus* is very close to *S. sibiricus,* and considerable difficulty is encountered in separating these two species. Perhaps the most obvious and constant difference is seen in the tendency of the context of the stipe base to stain vinaceous in *S. sibiricus.* As has already been indicated, it is now believed that *S. americanus* does not occur west of the Rocky Mountains and that many reports of its presence in the west have been based upon basidiocarps of *S. sibiricus.*

Edibility unknown.

SUILLUS SUBOLIVACEUS *Smith and Thiers, Contr. Toward a Monogr. of N. Am. Sp. of Suillus, p. 68. 1964*

Illustrations

Smith, A. H., and H. D. Thiers, A Contribution Toward a Monograph of North American Species of Suillus, pls. 24, 25.

Pileus 6.5–12 cm broad at maturity, convex when young, becoming plano-convex to plane to subumbonate at maturity; surface viscid, conspicuously appressed fibrillose, sometimes fibrillose-scaly on the margin, usually appearing glabrous and smooth on the disc; color brown ("cinnamon-brown" to "Verona brown") on the disc, fading to tan or light brown ("warm buff" to "buckthorn brown") toward the margin; margin decurved, becoming plane to uplifted, often lacerate with age. *Context* 1–2 cm thick, whitish to pale yellow, unchanging or becoming dingy pinkish ("avellaneous") when exposed. Taste and odor when fresh not distinctive.

Tubes 1–2 cm long, depressed, yellow ("straw yellow" to "Naples yellow" to "barium yellow"), unchanging or becoming brownish when exposed; *pores* 1–2 mm broad, angular, yellow ("primuline yellow" to "wax yellow"), sometimes becoming brown with age, unchanging or becoming brown when bruised.

Stipe 7–10 cm long, 1–1.5 cm thick at the apex, equal to subclavate, solid; surface dry, conspicuously glandulose, background color white to whitish, occasionally yellow at the apex; conspicuous, broad (up to 2 cm), gray to whitish annulus in apical region. *Context* white, unchanging when exposed.

Spore print brown. *Spores* 9–11 X 3–4 µm, subfusoid to subellipsoid, obscurely ventricose in side view, smooth, thin-walled.

Basidia 18–25 X 5–7 µm, clavate, hyaline, four-spored. *Hymenial cystidia* 40–50 X 6–9 µm, in clusters, incrusted, staining dark brown in KOH, thin-walled, subclavate to subfusoid.

Tube trama divergent, gelatinous, hyphae 2–4 µm wide. *Pileus trama* interwoven, homogeneous. *Pileus cuticle* differentiated as an ixotrichodermium, hyphae 3–5 µm wide. *Caulocystidia* similar to the cystidia in the hymenium. *Clamp connections* absent.

Chemical reactions KOH—cuticle dark brown to black, context lavender; $FeSO_4$—context grayish brown.

Habit, habitat, and distribution Gregarious to scattered in humus in mixed conifers. It is most probably associated with western white pine. *Suillus subolivaceus* has not yet been found within California, but is to be expected since it has been found in quantity in both Oregon and Washington, and since western white pine is widely distributed in the northern portion of the state.

Observations This *Suillus* appears most closely related to *S. intermedius* and *S. subluteus,* two eastern species that are not known west of the Rocky Mountains. It is distinguished by a well-developed, sheathing annulus, the usually olive to olive-yellow color of the pileus, the olivaceous to grayish-vinaceous cortex in the base of the stipe, and a distinctly sweet odor that develops as the basidiocarps are drying. A conspicuous mat of mycelium is frequently seen around the base of the stipe, as noted for species of *Leucopaxillus,* but not commonly seen in *Suillus* species.

Edibility unknown.

SUILLUS TOMENTOSUS *(Kauffman) Singer, Snell and Dick, Mycologia 51:570. 1959*

Boletus tomentosus Kauffman, Papers Mich. Acad. Sci. Arts and Letters 1:117. 1921.

Suillus hirtellus var. *mutans* (Peck) Snell, Lloydia 7:23. 1942.

Xerocomus lenticolor Snell and Dick, Mycologia 52:448. 1960.

Illustrations: See Microfiche No. 49

Smith, A. H., Mushroom Hunter's Field Guide, p. 86.

Smith, A. H., and H. D. Thiers, The Boletes of Michigan, pl. 13.

Smith, A. H., and H. D. Thiers, A Contribution Toward a Monograph of North American Species of Suillus, pls. 17, 18.

Snell, W. H., and E. A. Dick, The Boleti of Northeastern North America, pl. 22.

Pileus 6–12 cm broad at maturity, acutely convex to convex when young, becoming plano-convex to plane to highly irregular in outline; surface viscid, but not glutinous, conspicuously fibrillose-scaly to squarrose scaly when young, unchanging or becoming more or less appressed fibrillose-scaly to rarely subglabrous with age; when young background color yellow ("appricot yellow" to light "cadmium yellow"), changing to lighter yellow ("pinard yellow" to "empire yellow") with age, scales and fibrils often colored grayish ("olive" to "drab") to yellow ("Naples yellow"), darkening to dark yellow ("yellow ocher" to antimony yellow"), in higher elevations fibrils and scales characteristically become bright red ("jasper red" to "dragons-blood red"); margin incurved when young, becoming decurved to highly irregular and uneven with age, entire, often eroded with age. *Context* 1–2.5 cm thick, yellow ("massicot yellow" to "naphthalene yellow"), bluing irregularly when exposed. Taste and odor not distinctive.

Tubes 1–1.5 cm long, adnate to subdecurrent when young, becoming depressed with age, yellow ("naphthalene yellow"), bluing when exposed; *pores* up to 1 mm broad, angular, brown ("Verona brown" to "snuff brown") when young, changing to yellow ("olive lake" to "old gold") with age, bluing when bruised.

Stipe 6–11 cm long, 1–2.5 cm thick at the apex, equal to typically clavate, solid; surface dry to moist, usually giving a resinous feel to the fingers, noticeably glandulose, ± concolorous with the pileus except for the glands, which are darker; no annulus. *Context* yellow, bluing when exposed.

Spore print dark olive brown to brown. *Spores* 7–11 X 3.5–4.5 μm, pale ochraceous in KOH and Melzer's, smooth, thin-walled, subfusoid to subellipsoid in face view, slightly ventricose in side view.

Basidia 25–31 X 6–9 μm, clavate, four-spored, hyaline to pale yellow in KOH. *Hymenial cystidia* 33–45 X 6–12 μm, scattered to numerous, more abundant on the pores, clustered, hyaline, staining dark brown in KOH, clavate to cylindric.

Tube trama gelatinous in KOH, hyaline, divergent, hyphae 5–6 μm wide. *Pileus trama* interwoven, homogeneous. *Pileus cuticle* differentiated as a tangled ixotrichodermium with scattered fascicles of hyphae forming the scales. *Stipe cuticle* composed of caulocystidia similar to the cystidia of the hymenium. *Clamp connections* absent.

Chemical reactions HNO_3—cuticle pale pink, tubes pink; $FeSO_4$—cuticle, context, and tubes pale gray.

Habit, habitat, and distribution Scattered to gregarious in soil under beach and lodgepole pine. Usually abundant wherever these pines occur and characteristically fruits during the late fall and early winter.

Material studied El Dorado County: Thiers 20763. Humboldt County: Thiers 13979, 17747. Lassen County: Kowalski 2073. Madera County: Thiers 20841. Mariposa County: Thiers 21022. Mendocino County: Jensen 47; Largent 58; Motta 11, 34; Thiers 8164, 8300, 8781, 8788, 9247, 9249, 9300, 9308, 9468, 9634, 9744, 10639, 10699, 11072, 11858, 14603, 18115, 21516, 23060, 23079. Mono County: Thiers 20967. Nevada County: Sundberg 333; Thiers 13131, 13158. San Francisco County: Thiers 8572, 12089. San Mateo County: Thiers 7503. Shasta County: Thiers 11984. Sierra County: Sundberg 306, 352; Thiers 13863, 21137. Tuolumne County: Ammirati 604; Thiers 13271.

Observations This very distinctive *Suillus* is distinguished by the strongly developed, fibrillose-scaly surface of the pileus, the bluing of the context upon exposure, and the absence of a veil or annulus. Although the fibrillose scales are grayish to olivaceous when young, they frequently become red to bright red with age, at least in regions where the temperature becomes relatively low at night, such as the Sierra Nevada and other mountain ranges. Because of this color change, it is sometimes misidentified as *S. americanus*, but its larger size and the blue discoloration of the context make it readily distinct.

Suillus tomentosus can be distinguished from other *Suillus* species having a fibrillose-scaly cuticle, such as *S. fuscotomentosus* and *S. californicus*, by the much paler color of the fibrils and the color changes in the context. It is distinguished from the species in which the context changes to blue, such as *S. caerulescens*, *S. ponderosus*, and *S. reticulatus*, by the fibrillose-scaly cuticle, the absence of a veil, and the lack of reticulations on the surface of the stipe.

Edible, but of poor quality.

SUILLUS UMBONATUS *Dick and Snell, Mycologia 52:446. 1960*

Illustrations: See Microfiche No. 50

Smith, A. H., Mushroom Hunter's Field Guide, p. 85, pl. 51.

Smith, A. H., and H. D. Thiers, A Contribution Toward a Monograph of North American Species of Suillus, pls. 29–31.

Snell, W. H., and E. A. Dick, The Boleti of Northeastern North America, pl. 20.

Pileus 4–10 cm broad, convex to acutely convex when young, becoming broadly convex to umbonate to plane when older, occasionally becoming highly irregular and undulating with age; surface viscid, often appearing conspicuously streaked with age, often with spots or flecks on the margin; background color tan to pale yellow ("cream-buff" to "chamois" to "warm buff" to "ochraceous-buff"), sometimes somewhat darker, unchanging or becoming darker with age, occasionally dull pinkish brown ("avellaneous") to olivaceous near the margin, plaques concolorous or often reddish brown ("ochraceous-tawny" to "tawny" to "russet"); margin incurved and attached to stipe by partial veil when young, becoming appendiculate as cap expands, but usually glabrous at maturity. *Context* 1–2 cm thick, pale yellow ("cream color" to "massicot yellow") in young basidiocarps, not changing or becoming pale olive buff with age, changing to pinkish brown ("avellaneous") upon exposure. Taste unpleasant but not acid; odor mild.

Tubes up to 1 cm long, adnate, becoming subdecurrent to depressed; when young deep yellow ("mustard yellow" to "amber yellow"), becoming dark yellow ("primuline yellow" to "old gold") with age, unchanging or rarely with rust-colored ("ochraceous-tawny") stains when bruised; *pores* 2–3 mm long, 1 mm broad, radially arranged, compound, concolorous with tubes.

Stipe 3–7 cm long, 0.5–1 cm thick at the apex, equal or slightly enlarged or tapering toward the base, white mycelium at the base, solid; surface moist to subviscid, tan ("warm buff") above the annulus, unchanging or becoming pallid ("pale olive-buff") toward the base, often staining brown when handled; glandulae not apparent on the surface when young, but usually becoming apparent and yellow brown to brown with age; annulus viscid, pallid ("pinkish cinnamon"), subapical but sometimes appearing subperonate. *Context* pale yellow, unchanging, or when young changing to pale lavender ("avellaneous").

Spore print olive brown. *Spores* 7–11 X 3.5–4.5 μm, pale ochraceous in Melzer's and KOH, smooth, thin-walled, ellipsoid to subfusoid, obscurely ventricose in side view.

Basidia 20–25 X 6–8 μm, hyaline, four-spored, clavate. *Hymenial cystidia* 25–40 X 6–10 μm, in clusters, hyaline but staining dark brown in KOH, scattered to numerous, cylindric to clavate.

Tube trama gelatinous in KOH, divergent, hyaline. *Pileus trama* interwoven, homogeneous. *Pileus cuticle* differentiated as an ixotrichodermium, hyphae 3–6 μm wide. *Caulocystidia* similar to hymenial cystidia. *Clamp connections* absent.

Chemical reactions unknown.

Habit, habitat, and distribution Gregarious to cespitose in soil under lodgepole and beach pine. Collections of this species have been made almost annually in Jackson State Forest in Mendocino County where the basidiocarps appear among Bishop and beach pine mixed with tanbark oak and other hardwoods. In the Sierra Nevada and Cascade ranges it has been found only in forests composed of lodgepole pine, fir, and alder. Since lodgepole pine and its close relative, beach pine, are common to both habitats, it is assumed that this fungus forms mycorrhizal associations with these conifers.

Material studied El Dorado County: Thiers 20765, 20816, 20833. Humboldt County: Thiers 17612, 21381. Mendocino County: Thiers 8779, 8949, 9291, 9345, 9723, 11075, 14612, 21324. Sierra County: Thiers 13161, 13201, 13202, 13210, 21131, 21132, 21141.

Observations *Suillus umbonatus* is characterized by the presence of a distinct, viscid annulus that is colored pale watery brown, and by the development of olivaceous tones in the pileus with age; it is further distinguished by the failure of the context of the stipe base to become vinaceous when exposed. As has already been pointed out, there is considerable confusion surrounding this species and its relationship with *S. sibiricus* and *S. americanus. Suillus americanus,* apparently not present in this state, does not have the noticeably viscid annulus, has a differently colored pileus, and develops reddish scales or plaques on the pileus with age. *Suillus sibiricus* can be distinguished by the absence of an annulus, the much brighter and more intensely yellow pileus, and the vinaceous discoloration of the context of the base of the stipe.

Edibility not determined.

SUILLUS VOLCANALIS *Thiers, Madrono 19:158. 1967*

Illustrations: See Microfiche No. 51

Pileus 9–12 cm broad when mature, convex to irregular when young, becoming plano-convex to highly irregular with an undulating margin to plane with an uplifted margin when older; surface often irregularly pitted, appearing closely appressed fibrillose to streaked, viscid; color when young yellow ("primrose yellow") with numerous buff-colored areas ("ochraceous-buff" to "ochraceous-tawny"), when older frequently with some areas remaining yellow ("picric yellow") and others reddish brown ("vinaceous rufous"), or in some basidiocarps the background yellow to tan ("antimony yellow" to "warm buff") with irregular areas colored pinkish ("testaceous"), when very old red ("russet" to "tawny") with paler yellowish areas intermixed; margin strongly incurved when young, usually with a white to pallid, conspicuous roll of partial-veil tissue attached, becoming decurved and glabrous with age. *Context* 1–2 cm thick, white to pale yellow ("ivory yellow") when young, typically becoming bright yellow ("massicot yellow") in older pilei, unchanging when exposed. Taste and odor not distinctive.

Tubes 1–1.5 cm long, shallowly to deeply depressed during all stages of development, when young bright yellow ("straw yellow"), becoming dark yellow ("aniline yellow") when older, unchanging when exposed; *pores* less than 1 mm broad, angular, concolorous with tubes, unchanging when bruised.

Stipe 3–6 cm long, 1.5–3 cm thick at apex, equal to occasionally somewhat ventricose to flared at the base, solid; surface dry, appearing glabrous to smooth, glandulae apparent at least in the apex when very old; color when young white, except pale yellow at the apex, when older becoming dark yellow ("massicot yellow") at the apex and brown to rust colored ("buckthorn brown" to "ochraceous-tawny" to "tawny") toward the base; no annulus. *Context* white when young, yellowish with age, unchanging when exposed.

Spore print brown. *Spores* 6.6–10 X 3–4 μm, cylindric to subellipsoid, ventricose in side view, hyaline in KOH, pale yellow in Melzer's, smooth, thin-walled.

Basidia 25–31 X 7–11 μm, hyaline in KOH, clavate, four-spored. *Hymenial cystidia* 24–47 X 5–8 μm, rare to absent along sides of tubes, scattered to relatively abundant on pores, fascicled, no solitary cystidia seen, dark brown in KOH, thin-walled, bases of many appearing incrusted, cylindric to subclavate, rust brown ("ochraceous-tawny") in Melzer's.

Tube trama hyaline, divergent from a distinct mediostratum, appearing subgelatinous in KOH, hyphae 3–5 μm wide. *Pileus trama* with a compactly interwoven layer near tubes, which stains dark brown in KOH, surmounted by a layer of loosely interwoven, more or less homogeneous hyphae. *Pileus cuticle* composed of a narrow ixotrichodermium with hyphae that appear to be heavily incrusted and stain dark brown in KOH, surmounting a broad layer of interwoven hyphae 5–6 μm wide. *Stipe cuticle* interwoven, subgelatinous in KOH with occasional clusters of caulocystidia similar to those in the hymenium. *Clamp connections* not seen.

Chemical reactions unknown.

Habit, habitat, and distribution Gregarious to scattered (often buried) in cinders and humus under Jeffrey pine. This species was found by uncovering "humps" of soil or cinders in the vicinity of Butte Lake and the Cinder Cone in Lassen Volcanic National Park. This is one of the very few boletes that forms basidiocarps during the summer months.

Material studied Lassen County: Thiers 12777, 12800, 12915, 12930, 12931.

Observations Because of the general stature of the basidiocarp and the absence of glandulae on the stipe, *S. volcanalis* appears most closely related to *S. brevipes.* However, *S. volcanalis* can be distinguished by the well-developed, noticeable cottony margin of fibrillose tissue, the yellow to tawny color of the pileus, which often appears distinctly fibrillose rather than glabrous, the longer and broader spores, and the darkening of the stipe with age.

Edibility unknown.

TYLOPILUS KARSTEN

Rev. Mycol. 3:16. 1881

Pileus moderate to large, convex to plano-convex to plane; surface dry to moist, not truly viscid, glabrous to more commonly tomentose or fibrillose, pallid to tan or pale vinaceous to very dark brown or blackish, often becoming brown or blackish; margin entire; context pallid, often changing to blue when exposed, sometimes eventually becoming brown. Taste and odor usually mild. Tubes 1–2 cm long, white to buff to pale pink or vinaceous (dark reddish brown in *T. pseudoscaber*); pores angular, concolorous, often changing to another color when bruised. Stipe usually equal; surface more or less smooth or reticulate, glabrous to pruinose or rarely appressed fibrillose, no annulus or veil, white to pallid becoming brown to brownish vinaceous (dark sooty brown during all stages in *T. pseudoscaber*). Spores flesh color to vinaceous to dark reddish brown in deposit, hyaline to ochraceous tawny in KOH, subellipsoid to subfusoid to subcylindric, smooth. Hymenial cystidia solitary, hyaline, sometimes appearing incrusted, clavate to fusoid to fusoid-ventricose. Tube trama divergent. Cuticle a trichodermium or rarely a cutis. Clamp connections absent.

Tylopilus Karsten

Key to Species

1. Tubes yellow; many spores truncate and amyloid	*T. amylosporus*	
1. Tubes differently colored; spores not as above		2
2. Pores very dark brown to dark reddish brown; cap black to dark smoky brown	*T. pseudoscaber*	
2. Basidiocarp not as above		3
3. Surface of stipe reticulate	*T. indecisus*	
3. Surface of stipe not reticulate or only at the apex		4
4. Stipe often short, eccentric, or poorly developed; basidiocarp rarely coming fully above the ground	*T. humilis*	
4. Basidiocarp not as above		5
5. Pileus very dark, dull brown	*T. ferrugineus*	
5. Pileus pale vinaceous to tan or drab	*T. ammiratii*	

TYLOPILUS AMMIRATII *sp. nov.*

Pileus 5–10 cm latus, convexus vel pulvinatus, siccus, glaber, subbrunneus vel bubalinus, tactu fusco-brunneus. Contextus 1–1.5 cm crassus, albus vel avellaneus, immutabilis vel tarde brunnescens. Tubuli 0.8–1.5 cm longi, albi vel pallidi demum bubalini vel vinacei, tactu brunnei. Stipes 4–9 cm longus, 1.5–2.5 cm crassus, glaber vel pruinosus, non reticulatus, albus, tactu vinaceus vel brunneus. Sporae 8.6–11.4 X 3.3–4 µm, subfusoideae vel cylindricae. Cuticula intertexta. Holotypus (no. 6177) a Joseph Ammirati, Jr., lectus prope Mount Shasta, Shasta County, November 21, 1971; in Herbarium San Francisco State University conservatus.

Pileus 5–10 cm broad, convex to pulvinate, sometimes becoming undulating and uneven with age; surface dry to "sticky" but not truly viscid; color pale vinaceous-brown to buff ("pale purplish vinaceous" to "light vinaceous-fawn" to "pale vinaceous-pink" to "tilleul buff"), with age often drab ("vinaceous-drab" to "brownish drab"), brown to ochraceous colors appear on exposed pilei, changing to dark brown ("olive-brown" to "buffy brown" to "wood brown") when bruised, glabrous to occasionally obscurely velutinous; margin entire, sometimes with a slight sterile border. *Context* 1–1.5 cm thick, white to drab ("pale olive-buff" to "vinaceous-drab"), unchanging or sometimes slowly becoming pale brown when exposed. Taste and odor mild.

Tubes 0.8–1.5 cm long, depressed with decurrent lines extending down the stipe for 2–3 mm, white to pallid when young, becoming buff to vinaceous ("avellaneous" to "light vinaceous-fawn" to "vinaceous-buff") with age, changing to brown ("vinaceous-buff" to "wood brown") when bruised; *pores* one to two per millimeter, irregular to somewhat rounded, concolorous.

Stipe 4–9 cm long, 1.5–2.5 cm thick, equal to tapering downward, occasionally with a slightly enlarged base, solid; surface dry, glabrous to scurfy-pruinose, not reticulate, or only obscurely so for 1 cm at the apex resulting from decurrent tubes; color white, staining vinaceous to brown ("avellaneous" to "purple-drab" to "blackish purple") when bruised. *Context* white or mottled with pale brownish spots, changing to brown when exposed.

Spore print color not determined. *Spores* (7.5)8.6–11.4(13.6) X 3.3–4 µm, subfusoid to cylindric, inequilateral, hyaline to very pale ochraceous in KOH, ochraceous to dextrinoid in Melzer's, walls smooth, thin.

Basidia 20–25 X 5–8 µm, clavate, hyaline, staining cinnamon on the pores, four-spored. *Hymenial cystidia* 45–69 X 5–10 µm, scattered to numerous, often very obscure, clavate to irregular in outline, hyaline in KOH, bright rust ochraceous in Melzer's.

Tube trama divergent from a distinct central strand, gelatinous, hyaline except for the pale ochraceous color of the strand, hyphae 5–7 µm wide. *Pileus* *trama* interwoven, hyaline in KOH, homogeneous. *Pileus cuticle* differentiated as in interwoven to tangled trichodermium, no pileocystidia, ochraceous in KOH and Melzer's, hyphae 3–5 µm wide. *Stipe cuticle* interwoven with no basidia or caulocystidia. *Clamp connections* absent.

Chemical reactions KOH–cuticle brown to brownish ochraceous, context dull light yellow; $FeSO_4$–context dull bluish green.

Habit, habitat, and distribution Scattered to gregarious in soil under oaks. Known only from the Mount Shasta area in Shasta County.

Material studied Shasta County: Ammirati 6177 type; Thiers 28569 (collection made by Joseph Ammirati).

Observations This species belongs to the same complex as *T. indecisus* and *T. ferrugineus. Tylopilus ammiratii* appears much more closely related to *T. indecisus;* however, *T. indecisus* has a reticulate stipe, a considerably darker colored pileus and stipe, longer spores, and a fertile, cellular cuticle on the stipe. Both species occur under oaks, but *T. indecisus* is known only from under valley oaks, whereas *T. ammiratii* has been found only under California black oak. *Tylopilus ferrugineus* is distinguished by the very dark brown color of the pileus and the smaller pores.

Edibility unknown.

TYLOPILUS AMYLOSPORUS *(Smith) Smith. Non-Gilled Fleshy Fungi, p. 179. 1973*

Porphyrellus amylosporus Smith, Mycopathol. Mycol. Appl. 26:397. 1965.

Pileus 3–14 cm broad, convex when young, becoming broadly convex to plano-convex to plane with a decurved margin with age; surface dry, dull, velutinous to tomentose, typically becoming rimose with age and then sometimes appearing fibrillose or fibrillose-scaly; color dark olive brown to grayish brown ("olive-brown" to "citrine-drab" to "olive-buff"), unchanging or changing only slightly with age, cracks typically pallid ("warm buff" to "light buff"), no reddish pigmentation apparent; margin entire, incurved, becoming broadly decurved to plane. *Context* 0.5–1 cm thick, yellow to pallid ("pale olive-buff" to "warm buff"), erratically changing to blue when exposed, or sometimes unchanging. Taste and odor mild.

Tubes 1–2 cm long, depressed around the stipe, yellow ("Isabella color" to "olive-ocher" to "old gold"), staining blue when bruised; *pores* approximately 1 mm broad, angular, typically concolorous with the tubes or often with reddish discolorations, staining blue when bruised.

Stipe 5–10 cm long, 1–2 cm thick at apex, equal to tapering slightly toward the base, occasionally slightly flared at the apex, with yellow mycelium at the base, solid; surface dry, usually with small dots or flecks, but never truly fibrillose or tomentose, background color pale tan or pallid ("warm buff" to "pale olive-buff"), but this color often masked with reddish punctations or scales. *Context* yellow to occasionally red in midportion, unchanging when exposed.

Spore print dark reddish brown. *Spores* 13–17.6 X 4–6.5 µm, fusoid to subcylindric, often with considerable pleomorphism, amyloid, varying from only slightly so to (in one collection) strongly so, apex typically thickened and noticeably truncate, with a thin-walled depression in the center, walls smooth.

Basidia 30–35 X 8–12 µm, four-spored, clavate, hyaline. *Hymenial cystidia* 43–55 X 9–14 µm, scattered, hyaline, thin-walled, clavate with a tapering, subacute apex to near fusoid.

Tube trama obscurely divergent, hyaline in KOH, laticiferous hyphae usually present. *Pileus trama* interwoven, hyaline in KOH, laticiferous hyphae present. *Pileus cuticle* differentiated as a trichodermium of tangled hyphal tips, ochraceous in KOH, with somewhat irregularly incrusted, terminal cells 10–15 µm wide. *Clamp connections* not seen.

Chemical reactions unknown.

Habit, habitat, and distribution Solitary to scattered in humus in mixed northern coastal forests.

Material studied Mendocino County: Thiers 18511. San Mateo County: Thiers 23136, 27019. Santa Cruz County: Thiers 18547.

Observations This bolete is rather similar in appearance to members of the *Boletus truncatus–B. chrysenteron* complex, and in some cases can only be accurately distinguished from them by obtaining a good spore deposit. As indicated, the spores of *T. amylosporus* are dark reddish brown in mass, whereas in the others they are olive brown. The amyloid reaction of the spores of *T. amylosporus* often appears somewhat erratic and may vary from strongly positive to weakly so, or almost nonamyloid. In the *B. truncatus–B. chrysenteron* complex, however, the spores are consistently nonamyloid.

Edibility unknown.

TYLOPILUS FERRUGINEUS *(Frost) Singer. Am. Midland Nat. 37: 106. 1947*

Illustrations

Snell, W. H., and E. A. Dick, The Boleti of Northeastern North America, pl. 60.

Pileus 5–7 cm broad, obtusely convex to convex when young, typically becoming broadly convex to plane or shallowly depressed to highly irregular or undulating with age; surface dry, smooth, not rimose or areolate, tomentose to velutinous when young, often becoming glabrous with age; color dark brown ("Verona brown" to "snuff brown" to "warm sepia" to "bister") during all stages of development; margin incurved, becoming decurved, entire when young, often becoming somewhat lacerate with age. *Context* 1.5–2 cm thick, white to pallid ("pale pinkish buff" to "pale olive-buff"), unchanging with age or when exposed. Taste and odor mild.

Tubes 0.5–1 cm long, deeply depressed, flesh color ("pale cinnamon-pink" to "pale pinkish cinnamon" to "light pinkish cinnamon"), often with age assuming a yellowish cast, changing to brown when bruised; *pores* small, less than 1 mm broad, angular, concolorous, readily changing to brown when bruised.

Stipe 3.5–7 cm long, 1–2 cm thick at the apex, somewhat ventricose to equal or tapering sharply at the base, solid; surface dry, pruinose to obscurely farinose to occasionally fibrillose, in some, obscure reticulations on the apex; color white to pallid ("pale pinkish buff"), readily staining brown when bruised. *Context* white, staining brown when bruised or exposed.

Spore print color not determined. *Spores* 8–11 X 3–4 μm, hyaline in KOH, pale ochraceous in Melzer's, occasional spore dextrinoid, subcylindric to subfusoid, inequilateral, walls smooth, thin.

Basidia 24–28 X 5–8 μm, clavate, four-spored, hyaline in KOH and Melzer's. *Hymenial cystidia* 26–37 X 6–9 μm, clavate to cylindric to subfusoid, hyaline in KOH, dark ochraceous and appearing incrusted when mounted in Melzer's, thin-walled.

Tube trama obscurely divergent from a relatively broad central strand, hyaline in KOH, strand somewhat pale yellow in KOH, hyphae 3–5 μm wide. *Pileus trama* interwoven, homogeneous, hyaline in KOH. *Pileus cuticle* differentiated as a compactly interwoven trichodermium, hyaline to pale ochraceous in KOH, dark brown in Melzer's, hyphae 3–5 μm wide. *Clamp connections* absent.

Chemical reactions not determined.

Habit, habitat, and distribution Cespitose to gregarious in soil in mixed woods. Known only from a single collection made in the vicinity of Grass Valley and Nevada City.

Material studied Nevada County: Thiers 9683.

Observations The identification of this collection is somewhat tentative and was made with the use of Snell and Dick's manual on northeastern boletes. It seems to match their description and illustration. It is distinguished most readily by the very dark brown color of the pileus and stipe, the very firm consistency of the context, and the relatively small spores.

Edibility unknown.

TYLOPILUS HUMILIS *Thiers, Mycologia 58:815. 1966*

Illustrations; See Microfiche No. 52

Pileus 5–12 cm broad at maturity, convex to subglobose when young, becoming broadly convex to plano-convex to highly irregular with age, frequently with an undulating or convoluted margin, often not appearing fully expanded at maturity; surface dry, but sometimes viscid to subviscid in wet weather, dull to rarely shining or glistening when old, velutinous to subtomentose, often appearing matted-tomentose when mature, sometimes becoming areolate with age; color brown ("cinnamon-brown" to "Prout's brown" to "snuff brown" to "Verona brown" to "Saccardo's umber") during all stages of development; margin sterile, entire, incurved, unchanging or becoming decurved with age. *Context* 1–2 cm thick on the disc, firm, white to whitish, staining pale vinaceous ("avellaneous") when exposed (irregular areas of blue discoloration observed in one collection). Taste and odor mild, not distinctive.

Tubes 5–10 mm long, depressed, often narrowly and shallowly so, white when young, becoming pale flesh color ("tilleul buff" to pale "salmon-color" to "seashell pink" to "pinkish vinaceous"); *pores* one to two per millimeter, angular, concolorous with the tubes, staining brown ("cinnamon brown" to "Prout's brown") when bruised.

Stipe 2–5 cm long, 0.5–3.5 cm thick at maturity, characteristically short and sometimes poorly developed, subbulbous to frequently equal except for an abrupt pinching or tapering at the base, sometimes eccentric, solid; surface dry, glabrous, smooth, except frequently reticulate at the very apex, white or whitish at the apex, unchanging or sometimes becoming pale vinaceous ("avellaneous") toward the base when old, staining brown ("cinnamon-brown" to "Prout's brown") when bruised. *Context* white, staining pale vinaceous ("avellaneous"), except sometimes yellow in outer layer near the base when exposed (irregular bluish areas observed in one collection).

Spore print dull reddish brown. *Spores* 8–12 × 3–4 μm, hyaline in KOH, pale yellow in Melzer's, cylindric to subfusoid to subellipsoid, smooth, thin-walled.

Basidia 32–34 × 8–12 μm, clavate, four-spored, hyaline in KOH. Hymenium staining brown in KOH. *Hymenial cystidia* 30–66 × 7–12 μm, scattered to abundant, hyaline to yellow to occasionally dark brown in KOH, color sometimes fading, not incrusted, embedded in the hymenium, subcylindric to subfusoid to fusoid to fusoid-ventricose to clavate or mucronate with elongated apices.

Tube trama hyaline in KOH, strongly divergent from a distinct mediostratum that stains pale brown in KOH. *Pileus trama* homogeneous, interwoven. *Pileus cuticle* differentiated as a very thin layer of interwoven to appressed, nonincrusted hyphae that stain brown in KOH, underlain by a layer, 150–250 µm thick, of gelatinous to subgelatinous hyphae. *Stipe cuticle* differentiated as a layer of clavate cells, 20–35 X 6–10 µm, that stain brown in KOH, with scattered caulocystida 33–52 X 9–15 µm that are typically mucronate with tapering, subacute apices, staining dark brown in KOH, with a layer of hyaline, gelatinous to subgelatinous, interwoven hyphae underneath. *Clamp connections* absent.

Chemical reactions KOH—cuticle black; NH_4OH—cuticle red; HCl—cuticle pink; HNO_3—context negative to pale pink, cuticle bright pink to red; sulfoformalin—cuticle bright pink; guaiac—context dark blue; $FeSO_4$—context gray to green.

Habit, habitat, and distribution Solitary to gregarious to cespitose in a mixture of Bishop and beach pine. Fruiting only during the late fall and early winter season. This bolete certainly is one of the rarest of all the California species. It is known only from a single open grove of pines covering not more than half an acre in the vicinity of the now destroyed old dormitory of the fire station in Jackson State Forest east of Mendocino.

Material studied Mendocino County: Thiers 8391, 8815, 9009, 9067, 9290, 9333, 9347, 9437, 9687, 9813, 10582, 10594, 11053, 14157, 14599, 21276.

Observations The most unusual feature of this species is the apparent trend toward becoming gastroid. The stipe is frequently poorly developed and eccentric, and the tubes are sometimes permanently covered by the cuticle of the pileus. Furthermore, the basidiocarps often remain below ground until full maturity and some become only partially exposed. However, *T. humilis* is not a true puffball, since the spores are forcibly discharged.

Edibility unknown.

TYLOPILUS INDECISUS *(Peck) Murrill, Mycologia 1:15. 1909*

Boletus indecisus Peck, Ann. Rept. N.Y. State Mus. 41:76. 1888.

Illustrations: See Microfiche No. 53

Smith, A. H., and H. D. Thiers, The Boletes of Michigan, pl. 60.

Snell, W. H., and E. A. Dick, The Boleti of Northeastern North America, pl. 61.

Pileus 6–14 cm broad, convex to broadly convex when young, becoming plane to plano-convex to shallowly depressed on the disc with an undulating margin when mature, surface moist to dry, somewhat glistening but not viscid, becoming dull with age or when handled, cuticle firm, not soft, glabrous, rarely appearing subtomentose or velutinous; color dark brown ("hazel" to "Kaiser brown" to "russet" to "chestnut-brown") when young, becoming paler ("sayal brown" to "clay color" to "cinnamon-buff") with age, staining dark brown when bruised; margin incurved, becoming decurved, entire, becoming eroded. *Context* 1.5–3 cm thick, white when first exposed, unchanging or slowly darkening to pale brown ("clay color") near the cuticle and tan ("warm buff") elsewhere. Taste mild, sometimes slightly sweetish; odor mild.

Tubes 0.5–1 cm long, depressed, becoming somewhat decurrent with age, when young pink to pale flesh color ("avellaneous" to "cinnamon-buff" to "pinkish buff"), darkening slightly with age, staining brown when bruised or exposed; *pores* less than 1 mm broad, angular, concolorous with the tubes.

Stipe 8–10 cm long, 2–4.5 cm thick at the apex, equal to subclavate, white mycelium at the base, solid; surface dry, noticeably reticulate over three fourths of to entire length, buff to brown ("pinkish buff" to "cinnamon"), dark brown when bruised, reticulations concolorous or slightly darker than the surface. *Context* white when first exposed, changing to tan ("warm buff") in the apical region and vinaceous ("avellaneous" to "wood brown") toward the base.

Spore print flesh color. *Spores* 10.5–13 X 3–5 μm, hyaline in KOH, hyaline to pale ochraceous in Melzer's, rarely dextrinoid, narrowly cylindric to ellipsoid in face view, ventricose in profile, walls smooth, thin.

Basidia 18–22 X 7–5 μm, guttulate in KOH, hyaline, clavate, four-spored. *Hymenial cystidia* rare to scattered, inconspicuous, hyaline in KOH, reddish brown in Melzer's, clavate to fusoid-ventricose, sometimes appearing somewhat incrusted and then staining pale yellow in KOH.

Tube trama strongly divergent from a distinct central strand, hyaline, gelatinous in KOH, hyphae 6–8 μm wide. *Pileus trama* interwoven, homogeneous. *Pileus cuticle* a tangled trichodermium with scattered erect hyphae, terminal cells not differentiated, not incrusted, reddish brown in KOH, hypodermium well differentiated and subgelatinous in KOH. *Stipe cuticle* differentiated as a palisade of caulocystidia and basidioles, staining brown in KOH. *Clamp connections* absent.

Chemical reactions KOH–cuticle black; NH_4OH–cuticle red; HCl–cuticle yellow brown; $FeSO_4$–stipe dark gray, pileus context gray, cuticle black; guaiac–positive on all parts.

Habit, habitat, and distribution Scattered to gregarious in soil under live oaks in the coastal foothills. Known only from the Green Valley area in Solano County. Although relatively common in the midwestern and eastern United States, so far as is known this is the first report of its occurrence west of the Rocky Mountains.

Material studied Solano County: Thiers 24508, 26998.

Observations *Tylopilus indecisus* is easily distinguished from other *Tylopilus* species occurring within the state by the well-developed, expanded pileus, the strongly reticulate stipe, and the flesh-colored rather than dark-brown pores. The nonamyloid spores further distinguish it from *T. amylosporus.*

Edible.

TYLOPILUS PSEUDOSCABER *(Secretan) Smith and Thiers, Boletes of Michigan, p. 96. 1971*

Boletus pseudoscaber Secretan, Mycogr. Suisse 3:13. 1833.

Porphyrellus pseudoscaber (Secretan) Singer, Farlowia 2:115. 1945.

Boletus olivaceobrunneus Zeller and Bailey, Mycologia 27:457. 1935.

Porphyrellus atrofuscus Dick and Snell, Mycologia 52:449. 1960.

Illustrations: See Microfiche No. 54

Singer, R., Die Rohrlinge, Teil II, pl. XXVI, figs. 4–9.

Smith, A. H., and H. D. Thiers, The Boletes of Michigan, pl. 47.

Pileus 6–15 cm broad at maturity, convex to broadly convex when young, becoming plano-convex to irregular with an undulating to uneven margin with age; surface dull, unpolished, velutinous to tomentose during all stages of development, smooth, rarely finely areolate and fibrillose-scaly; color sooty black to very dark brown ("clove brown" to "fuscous" to "mummy brown"), occasionally slightly darker on the disc and paler on the margin, black when dried; margin entire, incurved to decurved. *Context* 1–2 cm thick, white, slowly staining weakly blue in some areas, pink ("testaceous") in other areas, blue discoloration slowly disappearing and becoming pink ("testaceous"), staining wax paper dark blue green. Taste not distinctive; odor often strong and pungent.

Tubes 1.5–2 cm long, shorter toward the margin, deeply and broadly to narrowly depressed, dark brown to almost black ("cinnamon-drab" to "drab" to "benzo brown"), slightly paler on the margin, staining blue at first when bruised, then changing to purplish brown ("deep brownish drab" to "dusky drab"); *pores* up to 1 mm wide with occasional larger pores irregularly distributed, angular, concolorous with the tubes, frequently bluing and then changing to dark mahogany brown when bruised.

Stipe 13–20 cm long, 1.5–2 cm thick at the apex, clavate to subclavate, solid to stuffed; surface dry, dull, glabrous, longitudinally ridged and often somewhat reticulate, dark brown ("bone brown" to "clove brown"), staining reddish vinaceous when first bruised, eventually becoming dark brown to black, occasionally scattered areas staining blue, whitish at the base. *Context* pale pinkish tan ("pale pinkish buff"), becoming pale flesh color ("testaceous") when exposed.

Spore print deep reddish brown. *Spores* 13.8–17.6 × 6–9.6 μm, dark ochraceous in Melzer's, pale yellow in KOH, smooth, thick-walled, subfusoid to subcylindric to subovoid in face view, inequilateral in side view, no apical pore.

Basidia 30–45 × 10–15 μm, clavate, four-spored, hyaline; irregular areas in hymenium staining dark brown in KOH. *Hymenial cystidia* 36–51 × 12–15 μm, abundant, cinnamon brown in KOH, chocolate brown in Melzer's, embedded in the hymenium, fusoid-ventricose to mucronate to clavate with elongated, tapering apices.

Tube trama divergent, hyaline, appearing somewhat gelatinous in KOH, hyphae ±8 μm wide. *Pileus trama* interwoven, hyaline except for scattered areas that stain brown in KOH, homogeneous. *Pileus cuticle* differentiated as a trichodermium, which frequently collapses with age, with scattered free hyphal tips colored cinnamon brown in KOH, not incrusted, contents of cells staining brown in KOH, hyphae up to 10 μm wide. *Stipe cuticle* differentiated as a tangled trichodermium that stains cinnamon brown in KOH. *Clamp connections* absent.

Chemical reactions KOH–cuticle dark red, context red; HNO_3, HCl, and H_2SO_4–context pink, cuticle bright red; sulfoformalin–context and cuticle red; $FeSO_4$–cuticle dark gray.

Habit, habitat, and distribution Scattered in the mixed coastal forests of northern California. It is often found growing on well-decomposed Sitka spruce logs and stumps, but has also been found in humus under Sitka spruce and various pines. Fruiting during the fall and winter season. Usually much more abundant in the area north of Eureka than elsewhere in the state.

Material studied Del Norte County: Thiers 17760. Humboldt County: Sundberg 977; Thiers 14248, 14355, 14417, 14443, 17642, 21222, 21389, 21462, 21484, 22889, 22904, 26883; White 508. Mendocino County: Orr 846.

Observations This very distinctive bolete is easily recognized by the dull-black color of the entire basidiocarp except for the basal portion of the stipe, which is often whitish. In addition, the context, at least in part, shows a blue discoloration when exposed, and in most instances wax paper will be stained blue to blue green when in contact with the basidiocarps. It is closely related to *T. porphyrosporus*, which has pores that are pallid to grayish in color and a context that does not change to blue when exposed.

Tylopilus pseudoscaber has frequently been misidentified on the west coast and has been the basis for the publication of at least two new species. Zeller and Bailey made collections of it in Oregon and described them as new under the name of *Boletus olivaceobrunneus.* Several years later Dick and Snell, studying basidiocarps from the vicinity of San Francisco, described them as new under the name of *Porphyrellus atrofuscus.* The type material of both of these species has been examined and no significant differences from *T. pseudoscaber* could be found.

Edibility unknown.

FIELD KEY TO CALIFORNIA BOLETES

GENERA

1. Tubes typically disoriented and irregularly arranged; spore deposit not obtainable	*Gastroboletus*
1. Tubes more or less vertically oriented and orderly arranged; spore deposit usually readily obtainable	2
2. Basidiocarps small (4–7 cm); tubes white when young, becoming bright yellow at maturity; spore deposit yellow; stipe typically hollow in the basal portion with age	*Gyroporus castaneus*
2. Basidiocarps typically larger; tubes yellow when young, or if white at first, then not bright yellow with age; spore deposit olivaceous to brown to reddish brown or flesh or vinaceous color; stipe usually not hollow	3
3. Basidiocarp with a conspicuous, cottony, bright yellow veil (be sure to check young specimens)	*Pulveroboletus ravenelii*
3. Basidiocarps lacking such a veil	4
4. Spore deposit flesh color to vinaceous to dark reddish brown; tubes at maturity eventually becoming pink to flesh color to vinaceous to dark reddish brown or even blackish (yellow in *Tylopilus amylosporus*)	*Tylopilus*
4. Spore deposit olivaceous to brown; tubes not colored as above	5
5. Cap viscid, subviscid, or dry; if dry, then stipe with an annulus; glandular dots usually apparent on stipe; if glandular dots lacking, then stipe annulate and pores radially arranged	*Suillus*

5. Cap dry to moist or rarely viscid; if viscid, stipe not annulate; glandular dots not present — 6
   6. Stipe with numerous small squamules, which are white or whitish when young but typically become black or brown with age; tubes white or whitish when young, becoming dull yellow or olivaceous with age — *Leccinum*
   6. Stipe glabrous, fibrillose, smooth, ridged or reticulate, dark colored squamules lacking; tubes yellow or white; if white, typically "stuffed" when young — *Boletus*

KEY TO GASTROBOLETUS

1. Context changing to blue when exposed or injured		2
1. Context unchanging or at least not changing to blue when exposed		3
2. "Cap" red to reddish brown; context changing to an intense blue immediately upon bruising	*Gastroboletus turbinatus*	
2. "Cap" brown to olive brown or yellow brown; context change to blue somewhat erratic and often slight	*Gastroboletus xerocomoides*	
3. "Tubes" covered with a white persistent peridial membrane that is distinct from the cuticle of the cap	*Gastroboletus subalpinus*	
3. Peridial membrane not formed		4
4. "Cap" brown, with no reddish tones or spots	*Gastroboletus suilloides*	
4. "Cap" buff to tawny, usually with reddish spots or areas irregularly distributed	*Gastroboletus amyloideus*	

KEY TO TYLOPILUS

1. Basidiocarp, including the tubes, blackish to fuscous to very dark reddish brown; tubes and cap staining wax paper blue green	*Tylopilus pseudoscaber*	
1. Basidiocarp not colored as above; wax paper not stained blue or blue green		2
2. Stipe strongly reticulate for one half to three quarters of length; cap pallid, tan to buff	*Tylopilus indecisus*	
2. Stipe not reticulate or for only a short distance at the apex; cap pallid to dark brown		3
3. Stipe short, poorly developed, often eccentric; sometimes basidiocarp not coming above ground; (known only from a single locality in Jackson State Forest near Mendocino in Mendocino County)	*Tylopilus humilis*	
3. Stipe typically centrally attached and well developed; basidiocarp coming above ground		4
4. Cap pale tan to pale vinaceous; tubes white, becoming flesh colored	*Tylopilus ammiratii*	
4. Cap dark brown to dark olive brown to gray brown; tubes flesh colored or yellow		5
5. Tubes yellow to yellowish at maturity; cap olive brown to gray brown	*Tylopilus amylosporus*	
5. Tubes flesh colored at maturity; cap dark brown	*Tylopilus ferrugineus*	

KEY TO SUILLUS

1. Cap surface dry, moist, or only subviscid	2
1. Cap surface viscid to glutinous	4
2. Cap surface dry, dull red to reddish brown, noticeably fibrillose-scaly	*Suillus lakei*
2. Cap surface moist to subviscid; brown to dark brown; fibrillose but not strongly fibrillose-scaly	3
3. Stipe surface noticeably reticulate	*Suillus reticulatus*
3. Stipe glandulose, not reticulate	*Suillus fuscotomentosus*
4. Stipe with a distinct annulus	5
4. Stipe lacking a distinct annulus	14
5. Context changing to blue, at least in base of stipe (change sometimes only slight and erratic)	6
5. Context unchanging or at least not changing to blue	9
6. Context of cap first becoming blue, then fuscous when exposed	*Suillus lithocarpi-sequoiae*
6. Context of cap usually unchanging; context in base of stipe changing to blue when handled or exposed but not becoming fuscous	7
7. Annulus heavy, touch, very viscid, lower surface orange; cap usually reddish brown, often with greenish stains when old	*Suillus ponderosus*
7. Annulus heavy, tough, very viscid, usually white or pallid; cap not colored as above	8
8. Cap glabrous, cinnamon to orange cinnamon; associated with spruce and true firs	*Suillus imitatus*
8. Cap with streaks or with scattered appressed fibrils; buff to pale vinaceous; associated with Douglas fir	*Suillus caerulescens*

9. Annulus well developed, often flaring or pendant		10
9. Annulus only developed as a fibrillose zone or ring around the stipe		11
10. Stipe with a well developed, peronate whitish veil forming an annulus that develops a purplish zone on the lower surface	*Suillus luteus*	
10. Stipe not peronate; annulus gray to white, not developing purple color on lower surface	*Suillus subolivaceus*	
11. Glands on surface of stipe obscure or not noticeable	*Suillus pseudobrevipes*	
11. Glands well developed and apparent during all stages of development		12
12. Cap yellow to bright yellow; stipe pallid (tan to buff); cap typically umbonate, at least when young	*Suillus umbonatus*	
12. Cap and stipe colored some shade of yellow; cap convex to plane		13
13. Cap bright yellow; stipe short, often somewhat eccentric	*Suillus megaporinus*	
13. Cap yellow brown; stipe well developed, not eccentric	*Suillus riparius*	
14. Context of stipe and cap changing to blue when exposed	*Suillus tomentosus*	
14. Context of cap and stipe not changing to blue when exposed		15
15. Cap margin with a noticeable cottony or fibrillose roll, at least when young		16
15. Cap margin glabrous or only with scattered fibrils		24
16. Pores large, up to 5 mm broad, often appearing somewhat lamellose; annulus sometimes present		17
16. Pores smaller, more or less angular or only slightly elongated; annulus not present		18
17. Cap bright yellow; stipe short, often somewhat eccentric	*Suillus megaporinus*	
17. Cap yellow brown; stipe well developed, not eccentric	*Suillus riparius*	

18. Cap white when young, then becoming gray and eventually reddish brown when mature; associated with knobcone and Monterey pines	*Suillus pungens*	
18. Cap not passing through the various color shades as given above		19
19. Cap white, becoming chocolate brown; associated with sugar pine	*Suillus brunnescens*	
19. Cap not colored as above; not associated with sugar pine		20
20. Cap dark cinnamon brown during all stages; stipe white becoming reddish brown	*Suillus borealis*	
20. Basidiocarp not colored as above		21
21. Pores 1–2 mm broad; cap dingy yellow; associated with white pine (*Pinus monticola*)	*Suillus sibiricus*	
21. Pores less than 1 mm broad; cap not colored as above; not associated with white pine		22
22. Glands conspicuous on stipe during all stages of development; associated with Bishop pine	*Suillus glandulosipes*	
22. Glands obscure, at least when young		23
23. Cap yellow to rust color when young, often spotted or mottled; associated with Jeffrey pine and possibly ponderosa pine	*Suillus volcanalis*	
23. Cap white to pallid to pale vinaceous when young; known only from under lodgepole pine in California	*Suillus albidipes*	
24. Cap surface with noticeable fibrils or fibrillose scales		25
24. Cap more or less glabrous or merely streaked		26
25. Cap "brown, pores large"; associated apparently with ponderosa pine (known only from a single collection made near Grass Valley, Calif. in 1914)	*Suillus californicus*	
25. Cap ochraceous to rust brown with dark fibrils or streaks; associated with Monterey and Bishop pines	*Suillus acerbus*	

26. Stipe 2–4 cm thick, clavate to ventricose	*Suillus monticolus*	
26. Stipe more or less equal, up to 2 cm thick		27
27. Stipe lacking conspicuous glands, at least when young	*Suillus brevipes*	
27. Stipe obviously glandular dotted during all stages		28
28. Tubes boletinoid (radiating from the stipe)	*Suillus punctatipes*	
28. Tubes not boletinoid	*Suillus granulatus*	

KEY TO LECCINUM

1. Cap some shade of orange to red to brown to maroon; margin with noticeable "flaps" of sterile tissue		2
1. Cap grayish to whitish; margin entire, lacking sterile "flaps" of tissue		10
2. Context of cap and stipe apex unchanging when exposed, or at least not becoming black or blue black	*Leccinum constans*	
2. Context eventually changing to black or blue black when exposed (change may be slow and erratic)		3
3. Context of cap and stipe apex changing to reddish or reddish brown before becoming blackish		4
3. Context of cap and stipe apex changing directly to blue black or blackish		6
4. Basidiocarps typically associated with madrones (*Arbutus*)	*Leccinum arbuticola*	
4. Basidiocarps typically associated with aspens (*Populus*)		5
5. Surface of cap matted fibrillose, at least when young, dry, rust red to apricot color	*Leccinum aurantiacum*	
5. Surface of cap glabrous when young, brown to rust red, dry to subviscid	*Leccinum discolor*	
6. Cap dark red to deep reddish brown; associated with madrone, manzanita, or toyon		7
6. Cap brown to reddish brown or pallid (pinkish), associated with conifers or hardwoods		8
7. Cap viscid during all stages; squamules on stipe coarse; pores white to pallid when young; associated with madrone or manzanita	*Leccinum manzanitae*	

7. Cap typically dry, becoming viscid only when wet for prolonged periods or when very old; squamules on stipe small and densely crowded; pores smoke colored; known only from under toyon — *Leccinum largentii*
   8. Cap brown or dull reddish brown, dark dull brown when dried — *Leccinum brunneum*
   8. Cap pallid (whitish to pale buff or pale pink) or orange to brick red — 9
9. Cap pallid (whitish to pale buff to pale pink); stipe white during all stages — *Leccinum armeniacum*
9. Cap orange to brick red; stipe white only when young — *Leccinum insigne*
   10. Cap gray to gray brown to blackish — *Leccinum montanum*
   10. Cap whitish to pale tan or tan — *Leccinum californicum*

KEY TO BOLETUS

Couplet	Species	Go to
1. Pores pink, red, or reddish brown		2
1. Pores white or yellow		8
2. Pores reddish brown, sometimes obscurely so; taste very acrid (biting)	*Boletus piperatus*	
2. Taste mild, or at least not acrid; pores red or pink		3
3. Surface of cap noticeably tomentose to velutinous or fibrillose; cap gray brown to near fuscous	*Boletus mendocinensis*	
3. Surface of cap glabrous or nearly so; cap not colored gray brown or fuscous		4
4. Stipe reticulate		5
4. Stipe not reticulate		6
5. Cap gray to pinkish gray; stipe conspicuously bulbous, pallid with pink overtones	*Boletus satanas*	
5. Cap brown to reddish brown; stipe clavate or only subbulbous, brown	*Boletus eastwoodiae*	
6. Cap bright, intense yellow	*Boletus orovillus*	
6. Cap differently colored		7
7. Cap pink to reddish vinaceous; pores pink	*Boletus amygdalinus*	
7. Cap dark brown to reddish brown; pores red	*Boletus erythropus*	
8. Cap viscid to glutinous; red to reddish brown; pores bright yellow	*Boletus flaviporus*	
8. Cap and tubes not as above		9
9. Stipe noticeably reticulate for at least one half the distance		10
9. Stipe not reticulate, or, if so, only at the very apex		17
10. Pores white when young, becoming yellow with age		11
10. Pores yellow during all stages of development		13

11. Surface of cap strongly reticulate or ridged; brown to dull yellow brown — *Boletus mottii*
11. Surface not as above — 12
    12. Cap pallid (tan to light brown); associated with pines — *Boletus edulis*
    12. Cap very dark brown to blackish; associated with oaks and madrone — *Boletus aereus*
13. Cap noticeably fibrillose to fibrillose-scaly, fibrils and scales colored dark brown — *Boletus fibrillosus*
13. Cap not as above — 14
    14. Cap deep rose to rose pink — *Boletus regius*
    14. Cap not colored as above — 15
15. Taste noticeably bitter — *Boletus calopus*
15. Taste mild, or at least not bitter — 16
    16. Cap glabrous; associated with oaks or other hardwoods — *Boletus appendiculatus*
    16. Cap often with scattered, appressed fibrillose scales; associated with firs at higher elevations — *Boletus abieticola*
17. Cap dark chocolate brown, surface conspicuously tomentose to fibrillose-scaly; stipe dark brown, sometimes alveolate at the apex — *Boletus mirabilis*
17. Cap and stipe not as above — 18
    18. Cap very dark gray brown to fuscous or blackish — 19
    18. Cap not colored as above — 21
19. Cap typically rimose or split or checked when old, at least near the margin — 20
19. Cap not becoming rimose — *Boletus zelleri*
    20. Cracks on cap shallow and exposed, context tan to pallid, not red — *Boletus truncatus*
    20. Cracks on cap deeper and exposed, context usually assuming reddish tints — *Boletus chrysenteron*
21. Cap red, reddish brown, or reddish with grayish overtones — 22
21. Cap not colored as above — 25

22. Cap red to dark reddish brown; surface glabrous to finely velutinous		23
22. Cap red to pinkish with grayish overtones; surface tomentose	*Boletus smithii*	
23. Stipe short, often flattened at the apex, colored red in the base and yellow in the apical portion; cap reddish brown	*Boletus dryophilus*	
23. Stipe not as above; cap not reddish brown		24
24. Stipe red, yellowish at the base; cap glabrous to fibrillose or velutinous	*Boletus amyloideus*	
24. Stipe buff to pale yellow, slight reddish blush sometimes at the apex; cap velutinous to pubescent	*Boletus coccyginus*	
25. Tubes changing to brick red when bruised (known only from two collections made near Stanford University several years ago)	*Boletus tomentipes*	
25. Tubes unchanging or, if changing, not becoming red when bruised		26
26. Cap olive brown; all parts of basidiocarp quickly and intensely changing to blue when bruised or exposed	*Boletus pulverulentus*	
26. Cap not colored as above; if changing to blue when bruised, change not intense or immediate		27
27. Cap tan to buff when young; stipe yellow at the apex, red at the base when young, entire stipe red when old	*Boletus rubripes*	
27. Basidiocarp not colored as above		28
28. Ammonium hydroxide giving a fleeting blue to blue-green color on the cap surface	*Boletus spadiceus*	
28. Ammonium hydroxide not giving a blue to blue-green reaction on the cap surface		29
29. Pores up to 3 mm broad	*Boletus subtomentosus*	
29. Pores up to 1 mm broad	*Boletus fragrans*	

BIBLIOGRAPHY

Chiu, W. F. 1948. The boletes of Yunnan. Mycologia 40:199–231.

Dick, E. A., and W. H. Snell. 1960. Notes on boletes: XIII. Mycologia 52: 444–454.

____. 1965. Notes on boletes: XV. Mycologia 62:448–458.

____. 1968. Some Boleti from Alaska. Mycologia 60:1204–1210.

Disbrey, B. D., and R. Watling. 1966. Histological and histochemical techniques applied to the Agaricales. Mycopathol. Mycol. Appl. 32:81–114.

Earle, F. S. 1904. Mycological studies: II. Bull. N.Y. Bot. Gard. 3:289–301.

Fries, E. 1821. Systema Mycologicum. I:1–508. Lund.

____. 1836–1838. Epicrisis Systema Mycologici. Upsala. 610 pp.

Groves, J. W., and S. C. Thomson. 1955. Notes on fungi from northern Canada. II: Boletaceae. Can. Field-Naturalist 69:44–51.

Harkness, H. W., and J. P. Moore. 1880. Catalogue of the Pacific Coast fungi. Calif. Acad. Sci. San Francisco. 46 pp.

Hoffman, T. C. 1963. A survey of the Boletaceae of Washington. Dissertation for the M.S. degree, University of Washington, Seattle.

Kallenbach, F. 1926–1938. Die rohrlinge in die pilze Mitteleuropas. I:1–158, 55 pls. Leipzig

Kauffman, C. H. 1925. The fungus flora of Mount Hood, with some new species. Papers Mich. Acad. Sci. Arts & Letters 5:115–148.

____. 1929. The fungous flora of the Siskiyou mountains in southern Oregon. Papers Mich. Acad. Sci. Arts & Letters 11:151–210.

Kawamura, S. 1966. Icones of Japanese fungi, Vol. 2. Kazamashobo, Tokyo. 283 pp.

Lange, J. E., and M. Lange, 1961. 600 pilze in farben. Munchen. 242 pp.

Lange, M., and F. B. Hora. 1963. A guide to mushrooms and toadstools. E. P. Dutton & Company, Inc., New York. 257 pp.

Leclair, A., and H. Essette. 1969. Les Bolets. Editions Paul Lechevalier. Paris. 81 pp.

McIlvaine, C. 1900. One-thousand American Fungi. Bowen-Merrill Co. Indianapolis. 704 pp., 182 pls.

Miller, O. K., and R. Watling. 1968. The status of Boletus calopus Fr. in North America. Notes Roy. Botan. Garden Edinburgh 28:317–326.

Murrill, W. A. 1910. Boletaceae. North Am. Flora 9:133–161.

____. 1912. Polyporaceae and Boletaceae of the Pacific Coast. Mycologia 4:91–100.

____. 1914. American boletes. Privately published. New York. 40 pp.

Pantidou, M. E. 1961. Culture studies of the Boletaceae: Gyrodon Merulioides and four species of Boletinus. Can. J. Botany 39:1149–1162.

Pearson, A. A. 1950. British Boleti. A. Brown & Sons Ltd. London. 17 pp.

Perreau-Bertrand, J. 1961. Recherches sur les ornementations sporales et la sporogeneses chez quelques especes des generes Boletellus et Strobilomyces (Basidiomycetes). Ann. Sci. Nat. Bot. 12e ser.:400–488.

Ridgeway, R. 1912. Color standards and color nomenclature. Published by the author. Washington, D. C. 53 pl.

Romagnesi, H. 1956–1967. Nouvel atlas des champignons. Vols. 1–4. Bordas. Paris.

Sharp, C. W., editor. 1963. Kitchen magic with mushrooms. The Mycological Society of San Francisco, Inc. San Francisco. 95 pp.

Singer, R. 1946. New and interesting species of Basidiomycetes. II. Papers Mich. Acad. Sci. Arts & Letters 32:103–151.

____. 1962. The Agaricales in modern taxonomy. 2nd ed. J. Cramer. Weinheim. 915 pp., 73 pls.

____. 1965. Die Rohrlinge, Teil I. J. Klinkhart. Bad Heilbrunn. 131 pp., 21 pls.

____. 1967. Die Rohrlinge, Teil II. J. Klinkhart. Bad Heilbrunn. 151 pp., 26 pls.

____. 1970. The topotype of Boletus amabilis. Mycologia 62:590–596.

Slipp, A. W., and W. H. Snell. 1944. Taxonomic–ecologic studies of the Boletaceae in northern Idaho and adjacent Washington. Lloydia 7:1–66.

____. 1944. Tentative keys to the Boletaceae of the Kaniksu National Forest. Northwest Sci. 17:17–24.

Smith, A. H. 1949. Mushrooms in their natural habitats. Sawyers, Inc. Portland. Ore. 626 pp.

____. 1958. The mushroom hunter's field guide, rev. ed. University of Michigan Press. Ann Arbor. 264 pp., 40 pls.

____. 1965. New and unusual basidiomycetes with comments on hyphal and spore wall reactions with Melzer's solution. Mycopathol. Mycol. Appl. 26:385–402.

Smith, A. H., and R. Singer. 1959. Studies on secotiaceous fungi. IV. Gastroboletus. Truncocolumella, and Chamonixia. Brittonia 11:205–223.

Smith, A. H., and H. D. Thiers. 1964. A contribution toward a monograph of North American species of Suillus. Privately published. Ann Arbor, Michigan. 116 pp., 46 pls.

____. 1966. Further notes on Suillus: the S. granulatus problem. Mycologia 58:469–474.

____. 1967. Comments on Suillus amabilis and Suillus lakei. Mycologia 59:361–367.

____. 1968. The generic position of Boletus subglabripes and Boletus chromapes. Mycologia 60:943–949.

____. 1968. A note on four species of Tylopilus. Mycologia 60:949–954.

____. 1968. Notes on the genus Suillus (Boletaceae). Mich. Botan. 7:14–18.

____. 1971. Boletes of Michigan. University of Michigan Press. Ann Arbor. 428 pp., 157 pls.

Smith, A. H., H. D. Thiers, and O. K. Miller. 1965. The species of Suillus and Fuscoboletinus of the Priest River Experimental Forest and vicinity, Priest River, Idaho. Lloydia 28:120–138.

Smith, A. H., H. D. Thiers, and R. Watling. 1966. A preliminary account of the North American species of Leccinum, section Leccinum. Mich. Botan. 5:131–179.

_____. 1967. A preliminary account of the North American species of Leccinum, sections Luteoscabra and Scabra. Mich. Botan. 6:107–154.

_____. 1968. Notes on species of Leccinum. I. Additions to section Leccinum. Lloydia 31:252–267.

Smith, H. V., and A. H. Smith. 1973. The Non-Gilled Fleshy Fungi. William C. Brown Company, Dubuque, Iowa. 402 pp.

Snell, W. H. 1934. Notes on boletes. III. Mycologia 26:348–359.

_____. 1936. Notes on boletes: IV. Mycologia 28:13–23.

_____. 1936. Notes on boletes: V. Mycologia 28:463–475.

_____. 1936. Tentative keys to the Boletaceae of the United States and Canada. Rhode Island Bot. Club 1:3–25.

_____. 1945. Notes on boletes: VII. Mycologia 37:374–388.

Snell, W. H., and E. A. Dick. 1941. Notes on boletes: VI. Mycologia 33:23–37.

_____. 1956. Notes on boletes: IX. Mycologia 48:302–310.

_____. 1958. Notes on boletes: X. A few miscellaneous discussions and a new subspecies. Mycologia 50:57–65.

_____. 1961. Notes on boletes: XIV. Mycologia 53:228–236.

_____. 1970. The Boleti of Northeastern North America. J. Cramer, Lehre. 115 pp., 87 pls.

Snell, W. H., R. Singer, and E. A. Dick. 1959. Notes on boletes. XI. Mycologia 51:564–577.

Stuntz, D. E., and B. F. Isaacs. 1962. Pacific Northwestern fungi. I. Mycologia 54:272–296.

Thiers, H. D. 1963. The bolete flora of the gulf coastal plain. I. The Strobilomycetae. J. Elisha Mitchell Sci. Soc. 79:32–41.

_____. 1964. The genus Xerocomus Quelet in northern California. Madrono 17:237–249.

_____. 1965. California boletes: I. Mycologia 57:524–534.

_____. 1966. California boletes: II. Mycologia 58:815–826.

_____. 1967. California boletes: III. The genus Suillus. Madrono 19:148–160.

_____. 1971. California boletes: IV. The genus Leccinum. Mycologia 63:261–276.

_____. 1971. Some ideas concerning the phylogeny and evolution of the boletes, pp. 423–440 in Evolution in the higher Basidiomycetes. University of Tennessee Press. Knoxville.

Thiers, H. D., and A. H. Smith. 1966. An undescribed bolete in the B. rubellus group. Mich. Botan. 5:117–118.

Thiers, H. D., and J. M. Trappe. 1969. Studies in the genus Gastroboletus. Brittonia 21:244–254.

Trappe, J. M. 1963. Some probable mycorrhizal associations in the Pacific Northwest. IV. Northwest Sci. 37:39–43.

Watling, R. 1970. British fungus flora. I. Boletaceae, Gomphidiaceae, Paxillaceae. Royal Botanical Garden. Edinburgh. 124 pp.

Whited, E. 1959. The black Boletus . . . a new discovery. Bull. Mycolog. Soc. San Francisco 9:9–11.

Zeller, S. M. 1914. The development of the carpophores of Ceriomyces zelleri. Mycologia 6:235–239.

_____. 1938. New or noteworthy agarics from the Pacific Coast states. Mycologia 30:468–474.

INDEX

Numbers in boldface type refer to page where species description begins